COSMIC EVOLUTION AND THE EVOLUTION OF LIFE

COSMIC EVOLUTION AND THE EVOLUTION OF LIFE

(A Hypothesis on the Metaphysics of Life)

Subir Bijoy Sengupta

KALPAZ

Cosmic Evolution and the Evolution of Life

Rs. 650

ISBN: 978-81-7835-896-3

Published in 2011 in India by
Kalpaz Publications
C-30, Satyawati Nagar,
Delhi-110052
E-mail: kalpaz@hotmail.com
Phone : 9212729499

Lasser Type Setting : Quick Media, Delhi
Printed at : Chawla Offset Printers, Delhi

"Philosophy is the formal embodiment of the 'penetrative imagination'; it deals with the significance of things; and transforms them, but only by intensified illumination".

—Bernard Bosanquet

CONTENTS

Preface 9

1. Primordial Stuff 11
2. Concepts of Time and Causality 35
3. Concept of Space and its Evolution—Concept of Energy and the Evolution of Matter 51
4. The Forces that Work on Matter—Evolution of Life and Mind 71
5. Principles behind the Evolution of the World 105
6. Activities of Men—Economic, Artistic, Philosophic and Religious 141
7. Struggles in Life and a Note on the Moral Life of Men 167
8. Knowledge and Its Origin 193
9. Relevance of Religion to Our Hypothesis 221

Index 247

PREFACE

If philosophy is an attempt to explain and appreciate life and the universe as a whole, then to explain life is a basic task of philosophers. The riddle of life has therefore drawn the attention of philosophers and scientists. Various explanations for the emergence of life have come out of their work and various theories have been recorded in our history of thought. But no such theory has yet been accepted by all and no theory has as yet explained all the aspects of this mysterious principle, life. The principle of life has eluded the philosophers and scientists perhaps because of the fact that by its very nature life is a mystical phenomenon. Most of the explanations advanced by thinkers have therefore been mystical, drawing no conclusions in clear terms. As living beings we are very much acquainted with life but to us it is still a principle shrouded in mystery only. Life is still a baffling problem for us. Even after our philosophical quest for a few thousand years life is still forcing us to use only figurative language to explain it. We are yet to find answers to the questions—What is life? How does it evolve and why?

We have in this work made an humble yet serious attempt to provide answers to the above questions. We have tried to enunciate in our hypothesis the principle that works behind the evolution of life in this universe. The main issue of our hypothesis being the evolution of life, we were required to work out the first principle or primordial stuff from out of which life has evolved. From this first principle we have derived the principles behind our concepts of time and causality and have worked out the principle behind the evolution of space and energy. We have then shown how matter evolves out of our first principle and occupies space. We have analysed the forces that work on matter and have come to the core of our hypothesis—the evolution of life. We have shown how the evolution of life is a logical consequence and result of the forces that work on matter. We have derived the principle of mind from out of life and have determined its ultimate destiny. We have then formulated the principles that work behind the evolution of this world. Now, man yet occupies the highest position in the scale of evolution. We have therefore sought to determine the position of men in this universe and have analysed their activities, indicating their ultimate destiny. We have also worked out our stand on the question of knowledge and its origin. Lastly, in order to make our hypothesis a complete system we have formulated the principles of morality in the light of our views on the evolution of this universe. We have then concluded our work showing the relevance of religion to our hypothesis.

As we have worked out our principles in outlines only our analysis may appear to be naïve, crude and an oversimplification of facts. But if simple concepts are enough to analyse our problem, what justification could we have to complicate the issue? If our method of deriving the basic principles seems to be like that of a magician who creates articles out of nothing, we shall beg to be excused and ask the reader to go through our text again. There are statements repeated over and over again in almost every chapter. We had to adopt this crude method as only a means to keep our basic stand clear and alive. If the line of our approach to the problem adds a new stream to the currents of philosophical thought we hope to be excused for our staccato language which might, on occasion, have failed sufficiently to clear up the issue.

SUSMITA, MOTINAGAR
SHILLONG, INDIA

Subir Bijoy Sengupta

Chapter 1
Primordial Stuff

Henri Bergson makes an elaborate enquiry into the different states of our consciousness and observes: "The greater part of the time we live outside ourselves, hardly perceiving anything of ourselves but our own ghost, a colourless shadow which pure duration projects into homogeneous space. Hence our life unfolds in space rather than in time; we live for the external world rather than for ourselves; we speak rather than think; we "are acted" rather than act ourselves".[1] Bergson shows that there are two different phases of our mental life corresponding to the activities of our superficial and fundamental selves. He writes, "There is at least one reality which we all seize from within, by intuition and not by simple analysis. It is our own person in its flowing through time, the self which endures. With no other thing can we sympathize intellectually, or if you like, spiritually. But one thing is sure: we sympathize with ourselves.

When, with the inner regard of my consciousness, I examine my person in passivity, like some superficial encrustment, first I perceive all the perceptions which come to it from the material world. These perceptions are clear-cut, distinct, juxtaposed or mutually juxtaposable; they seek to group themselves into objects. Next I perceive memories more or less adherent to these perceptions and which serve to interpret them; these memories are, so to speak, as if detached from the depth of my person and drawn to the periphery by perceptions resembling them, they are fastened on me without being absolutely myself. And finally, I become aware of tendencies, motor habits, a crowd of virtual actions more or less solidly bound to those perceptions and these memories. All these elements with their well-defined forms appear to me to be all the more distinct from myself the more they are distinct from one another. Turned outwards from within, together they constitute the surface of a sphere which tends to expand and lose itself in the external world. But if I pull myself in from the periphery toward the centre, if I seek deep down within me what is the most uniformly, the most constantly and durably myself, I find something altogether different.

What I find beneath these clear-cut crystals and this superficial congelation is a continuity of flow comparable to no other flowing I have ever seen. It is a succession of states each one of which announces what follows and contains what precedes. Strictly speaking they do not constitute multiple states until I have already got beyond them, and turn around to observe their trail. While I was experiencing them they were so solidly organized, so profoundly animated with a common life, that I could never have said where any one of them finished or the next one began. In reality, none of them do begin or end; they all dove-tail into one another".[2]

Our fundamental self is the self which endures and "Pure duration" in the language of Bergson, "is the form which the succession of our conscious states assumes when our ego lets itself live, when it refrains from separating its present state from its former states. For this purpose it need not be entirely absorbed in the passing sensation or idea; for then, on the contrary, it would no longer endure. Nor need it forget its former states: it is enough that, in recalling these states, it does not set them alongside its actual state as one point alongside another, but forms both the past and the present states into an organic whole, as happens when we recall the notes of a tune, melting, so to speak, into one another."[3] Bergson clarifies his position further when he writes, "We should therefore distinguish two forms of multiplicity, two very different ways of regarding duration, two aspects of conscious life. Below homogeneous duration, which is the extensive symbol of true duration, a close psychological analysis distinguishes a duration whose heterogeneous moments permeate one another; below the numerical multiplicity of conscious states, a qualitative multiplicity; below the self with well-defined states, a self in which succeeding each other means melting into one another and forming an organic whole. But we are generally content with the first, i.e., with the shadow of the self projected into homogeneous space. Consciousness, goaded by an insatiable desire to separate, substitutes the symbol for the reality, or perceives the reality only through the symbol. As the self thus refracted, and thereby broken to pieces, is much better adapted to the requirements of social life in general and language in particular, consciousness prefers it, and gradually loses sight of the fundamental self.

In order to recover this fundamental self, as the unsophisticated consciousness would perceive it, a vigorous effort of analysis is necessary, which will isolate the fluid inner states from their image, first refracted, then solidified in homogeneous space. In other words, our perceptions, sensations, emotions and ideas occur under two aspects: the one clear and precise, but impersonal; the other confused, ever changing, and inexpressible, because language cannot get hold of it without arresting its mobility or fit it into its common-place forms without making it into public property. If we have been led to distinguish two forms of multiplicity, two forms of duration, we must expect each conscious state, taken by itself, to assume a different aspect according as we consider it within a discrete multiplicity or a confused multiplicity, in the time as quality, in which it is produced, or in the time as quantity, into which it is projected."[4]

"What is duration within us?" Bergson continues, "A qualitative multiplicity, with no likeness to number; an organic evolution which is yet not an increasing quantity; a pure heterogeneity within which there are no distinct qualities. In a word, the moments of inner duration are not external to one another.

What duration is there existing outside us? The present only, or, if we prefer the expression, simultaneity. No doubt external things change, but their moments do not succeed one another, if we retain the ordinary meaning of the word, except for a consciousness which keeps them in mind. We observe outside us at a given moment a whole system of simultaneous positions; of the simultaneities which have preceded them nothing remains. To put duration in space is really to contradict oneself and place succession within simultaneity. Hence we must not say that external things endure, but rather that there is in them some inexpressible reason in virtue of which we cannot examine them at successive moments of our own duration without observing that they have changed. But this change does not involve succession unless the word is taken in a new meaning: on this point we have noted the agreement of science and common sense.

Thus in consciousness we find states which succeed, without being distinguished from one another; and in space simultaneities which, without succeeding, are distinguished from one another, in the sense that one has ceased to exist when the other appears. Outside us, mutual externality without succession; within us, succession without mutual externality".[5]

As regards space into which we project our feelings at the cost of losing the mobile character of time, Bergson writes, "Now, if space is to be defined as the homogeneous, it seems that inversely every homogeneous and unbounded medium will be space. For, homogeneity here consisting in the absence of every quality, it is hard to see how two forms of the homogeneous could be distinguished from one another. Nevertheless, it is generally agreed to regard time as an unbounded medium, different from space but homogeneous like the latter: the homogeneous is thus supposed to take two forms, according as its contents co-exist or follow one another.

It is true that, when we make time a homogeneous medium in which conscious states unfold themselves, we take it to be given all at once, which amounts to saying that we abstract it from duration. This simple consideration ought to warn us that we are thus unwittingly falling back upon space, and really giving up time. Moreover, we can understand that material objects, being exterior to one another and to ourselves, derive both exteriorities from the homogeneity of a medium which inserts intervals between them and sets off their outlines: but states of consciousness, even when successive, permeate one another, and in the simplest of them the whole soul is reflected. We may therefore surmise that time, conceived under the form of a homogeneous medium, is some spurious concept, due to the trespassing of the idea of space upon the field of pure consciousness. At any rate we cannot finally admit two forms of the homogeneous, time and space, without first seeking whether one of them cannot be reduced to the other. Now, externality is the distinguishing mark of things which occupy space, while states of consciousness are not essentially external to one another, and become so only by being spread out in time, regarded as a homogeneous medium. If, then, one of these two supposed forms of the homogeneous, namely time and space, is derived from the other, we can surmise a priori that the idea of space is the fundamental datum. But, misled by the apparent simplicity of the idea of time, the philosophers who have tried to reduce one of these ideas to the other have thought that they could make extensity out of duration. While showing how they have been misled, we shall see that time, conceived under the form of an unbounded and homogeneous medium, is nothing but the ghost of space haunting the reflective consciousness".[6] And then, "it follows from this analysis that space alone is homogeneous, that objects in space form a distinct multiplicity, and that every discrete multiplicity is got by a process of unfolding in space. It also follows that there is neither duration nor even succession in space, if we give to these words the meaning in which consciousness takes them: each of the so-called successive states of the external world exists alone; their multiplicity is real only for a consciousness that can first retain them and then set them side by side by externalizing them in relation to one another. If it retains them, it is because these distinct states of the external world give rise to states of consciousness which permeate one another, imperceptibly organize themselves into a whole, and bend the past to the present by this very process of connexion. If it externalizes them in relation to one another, the reason is that thinking of their radical distinctness (the one having ceased to be when the other appears on the scene), it perceives them under the form of a discrete multiplicity, which amounts to setting them out in line, in the space in which each of them existed separately. The space employed for the purpose is just that which is called homogeneous time."[7] Finally, "in a word, our ego comes in contact with the external world at its surface; our successive sensations, although dissolving into one another, retain something of the mutual externality which belongs to their objective causes; and thus our superficial psychic life comes to be pictured without any great effort as set out in a homogeneous medium. But the symbolical character of such a picture becomes more striking as we advance further into the depths of consciousness: the deep-seated self which ponders and decides, which heats and blazes up, is a self whose states and changes permeate one another and undergo a deep alteration as soon as we separate them from one another in order to set them out in space".[8]

As regards our preference for living with the superficial self, Bergson writes, "Our mind, which seeks solid bases of operation, has as its principal function, in the ordinary course of life, to imagine states and things. Now and then it takes quasi-instantaneous views of the undivided mobility of the real. It thus obtains sensations and ideas. By that means it substitutes for the continuous the discontinuous, for mobility stability, for the tendency in process of change it substitutes fixed points which mark a direction of change and tendency.

This substitution is necessary to common sense, to language, to practical life, and even, to a certain extent which we shall try to determine, to positive science. Our intelligence, when it follows its natural inclination, proceeds by solid perceptions on the one hand and by stable conceptions on the other".[9] And again, "The reason is that our outer and, so to speak, social life is more practically important to us than our inner and individual existence. We instinctively tend to solidify our impressions in order to express them in language".[10]

We recover our fundamental self only when our feelings are intense. In Bergson's language, "This overwhelming of the immediate consciousness is nowhere so striking as in the case of our feelings. A violent love or a deep melancholy takes possession of our soul: here we feel a thousand different elements which dissolve into and permeate one another without any precise outlines, without the least tendency to externalize themselves in relation to one another; hence their originality. We distort them as soon as we distinguish a numerical multiplicity in their confused mass: what will it be, then, when we set them out, isolated from one another, in this homogeneous medium which may be called either time or space, whichever you prefer? A moment ago each of them was borrowing an indefinite colour from its surroundings: now we have it colourless, and ready to accept a name. The feeling itself is a being which lives and develops and is therefore constantly changing; otherwise how could it gradually lead us to form a resolution? Our resolution would be immediately taken. But it lives because the duration in which it develops is a duration whose moments permeate one another. By separating these moments from each other, by spreading out time in space, we have caused this feeling to lose its life and its colour. Hence, we are now standing before our own shadow: we believe that we have analysed our feeling, while we have really replaced it by a juxtaposition of lifeless states which can be translated into words, and each of which constitutes the common element, the impersonal residue, of the impressions felt in a given case by the whole of society".[11] Bergson clarifies his position further when he writes, "Hence there are finally two different selves, one of which is, as it were, the external projection of the other, its spatial and, so to speak, social representation. We reach the former by deep introspection, which leads us to grasp our inner states as living things, constantly becoming, as states not amenable to measure, which permeate one another and of which the succession in duration has nothing in common with juxtaposition in homogeneous space. But the moments at which we thus grasp ourselves are rare, and that is just why we are rarely free. The greater part of the time we live outside ourselves, hardly perceiving anything of ourselves but our own ghost, a colourless shadow which pure duration projects into homogeneous space. Hence our life unfolds in space rather than in time; we live for the external world rather than for ourselves; we speak rather than think; we "are acted" rather than act ourselves. To act freely is to recover possession of oneself, and to get back into pure duration".[12]

On the metaphysics of matter Bergson writes, "Matter, in our view, is an aggregate of 'images'. And by 'image' we mean a certain existence which is more than that which the idealist calls a representation, but less than that which the realist calls a thing, an existence placed half-way between the 'thing' and the 'representation'. This conception of matter is simply that of common sense. It would greatly astonish a man unaware of the speculations of philosophy if we told him that the object before him, which he sees and touches, exists only in his mind and for his mind, or even, more generally, exists only for mind, as Berkeley held. Such a man would always maintain that the object exists independently of the consciousness which perceives it. But, on the other hand, we should astonish him quite as much by telling him that the object is entirely different from that which is perceived in it, that it has neither the colour ascribed to it by the eye, nor the resistance found in it by the hand. The colour, the resistance, are, for him, in the object: they are not states of our mind; they are part and parcel of an existence really independent of our own. For common sense, then, the object exists in itself, and, on the other hand, the object is, in itself, pictorial, as we perceive it: image it is, but a self-existing image.

We place ourselves at the point of view of a mind unaware of the disputes between philosophers. Such a mind, would naturally believe that matter exists just as it is perceived; and, since it is perceived as an image, the mind would make of it, in itself, an image. In a word, we consider matter before the dissociation which idealism and realism have brought about between its existence and its appearance."[13] And then, "Homogeneous space and homogeneous time are then neither properties of things nor essential conditions of our faculty of knowing them: they express, in an abstract form, the double work of solidification and of division which we effect on the moving continuity of the real in order to obtain there a fulcrum for our action, in order to fix within it starting-points for our operation, in short, to introduce into it real changes. They are the diagrammatic design of our eventual action upon matter."[14]

Bergson summarises the principles and processes of workings of our mental life thus: "Behind 'spirituality' on the one hand, and 'materiality' with intellectuality on the other, there are then two processes opposite in their direction, and we pass from the first to the second by way of inversion, or perhaps even by simple interruption, if it is true that inversion and interruption are two terms which in this case must be held to be synonymous, as we shall show at more length later on. This presumption is confirmed when we consider things from the point of view of extension, and no longer from that of duration alone.

The more we succeed in making ourselves conscious of our progress in pure duration, the more we feel the different parts of our being enter into each other, and our whole personality concentrate itself in a point, or rather a sharp edge, pressed against the future and cutting into it unceasingly. It is in this that life and action are free. But suppose we let ourselves go and, instead of acting, dream. At once the self is scattered; our past, which till then was gathered together into the indivisible impulsion it communicated to us, is broken up into the thousand recollections made external to one another. They give up interpenetrating in the degree that they become fixed. Our personality thus descends in the direction of space. It coasts around it continually in sensation."[15] Bergson continues, "In order that our consciousness shall coincide, with something of its principle, it must detach itself from the already-made and attach itself to the being-made. It needs that, turning back on itself and twisting on itself, the faculty of seeing should be made to be one with the act of willing—a painful effort which we can make suddenly, doing violence to our nature, but cannot sustain more than a few moments. In free action, when we contract our whole being in order to thrust it forward, we have the more or less clear consciousness of motives and of impelling forces, and even, at rare moments, of the becoming by which they are organized into an act: but the pure willing, the current that runs through this matter, communicating life to it, is a thing which we hardly feel, which at most we brush lightly as it passes."[16] And again, "Extension, we said, appears only as a tension which is interrupted. Or, are we considering the concrete reality that fills this extension? The order which reigns there, and which is manifested by the laws of nature, is an order which must be born of itself when the inverse order is suppressed; a detention inverse of the will would produce precisely this suppression. Lastly, we find that the direction, which this reality takes, suggests to us the idea of a thing unmaking itself; such, no doubt, is one of the essential characters of materiality."[17] Then, "Where the understanding, working on the image supposed to be fixed of the progressing action, shows us parts infinitely manifold and an order infinitely well contrived, we catch a glimpse of a simple process, an action which is making itself across an action of the same kind which is unmaking itself, like the fiery path torn by the last rocket of a fire-works display through the black cinders of the spent rockets that are falling dead."[18] "Intuition and intellect represent two opposite directions of the work of consciousness: intuition goes in the very direction of life, intellect goes in the inverse direction, and thus finds itself naturally in accordance with the movement of matter".[19] And finally, "If our analysis is correct, it is consciousness, or rather supra-consciousness, that is at the origin of life. Consciousness, or supra-consciousness, is the name for the rocket whose extinguished fragments fall back as matter; consciousness, again, is the name for that which subsists of the rocket itself, passing through the fragments and lighting them up into organisms."[20]

It appears that according to Bergson, there are two aspects of our mind — a superficial and a deeper one. The superficial aspect of our mental life consists of sensations, emotions, thoughts, etc. which are separated, isolated, named and arranged one after another. The element of time appears in this aspect as an order of succession because here the different states of mind appear one after another. But in the deeper aspect of mind, our experiences are inseparably intermingled. Our experiences here inter-penetrate each other and form a concrete whole. As such our experiences here cannot be arranged in any order of succession. The intense experiences like a violent love, a deep melancholy, etc. exhibit the deeper aspect of our mind and make us feel our existence in entirety. The moving character of such experiences shows that they are in time. But the time of such a feeling is devoid of the order of succession in it. Time here means the continuous development of a multiplicity of elements constantly running into one another, the process of unification going hand in hand with that of multiplication.

As regards matter that lie outside our mind, Bergson is in agreement with the view of Berkeley that it is an aggregate of images, colour, shape, etc. But the images constituting matter are, according to Bergson, independent of our mind. They are independent also of the perceiving act involved. The images are independent of our mind and the perceiving act because the fixed laws operating on them are beyond our control. Matter is just what appears to our senses, and set out in space. It is there where it is perceived. But at the root of images that appear as matter we feel a movement, which is a complex undifferentiated unity of feeling. This feeling explodes into a series of discrete and successive images that appear as matter to us. Thus at the root of matter, as we view it, there is an undifferentiated psychic unity. This unity is like a tension that gives way to a relaxation, or extension, or in the language of Bergson, a detention forming the objects that appear outside our mind. As there is movement in the said tension, so it must be in time. But as the various elements of such tension lie undifferentiated, there is no succession in the time here. When the tension is relaxed its elements appear in an order of succession or exhibit time as an order of succession.

We thus find that there are two aspects each of mind, matter and time. Mind, in its deeper aspect, is a tension of intermingled experiences and in its superficial aspect, mind consists of isolated elements of sensations, emotions, thoughts, etc. Similarly, matter, at its root, is an undifferentiated psychic unity or a tension in the mind. In its outer aspect, matter is a differentiated series of successive images set out in space. And time, in its inner aspect, is a duration that we feel in the tension of our mind where there is a continuous development of multiplicity of elements running constantly into one another in a process of unification going hand in hand with multiplication. Time, in its outer aspect, is an order of succession that is involved in the development of differentiated entities, mental and extra-mental.

The deeper aspect of our mind, which is a tension of intermingled experiences, contains in it the root of images that appears later on as matter and also time as duration.

This tension of our mind produces by detention the superficial aspect of mind, matter as a series of images, and time as an order of succession. The whole process is like the explosion of a rocket or the unwinding of a spring. The mind relaxes its tension for the purposes of practical life. Life requires for its success a pointed attention on particular aspects of its experiences. The exigencies of practical life require, therefore, the breaking up of its unified experiences into fragments and their treatment as discrete states or objects. A deeper experience of one's life, which is a complex, continuous and unbroken flux thus turns into discrete and isolated states and images outside one another.

While mind in its basic nature is a tension of unified experiences, matter is a relaxed form of such tension having differentiation and diversification in it. Mind and matter thus appear to us as radically different entities, though in reality they are not so. As principles of activity both mind and matter come under the same category. Mind is the forward tendency of creation and matter is creation that uncreates itself. We may, therefore, surmise that mind and matter are derived from a common source or rather from a basic principle. This principle is a tension of undifferentiated nature and it lets go its tension to create mind and matter. To get to this principle we are to plunge back into 'pure willing' or the pure impulse from which our narrowed consciousness and perception of matter emerge. As this impulse is a principle of creative activity and as it implies the primary functions of memory and anticipation, that is choice, which means the utilization of the past for the creation of a new future, we may call it 'consciousness', because 'consciousness' connotes such activities. The word 'consciousness' must not, of course, be taken in the sense of the narrowed consciousness that functions in each of us. 'Consciousness' here means the pure impulse of creative activity that produces our narrowed consciousness and matter. 'Consciousness' or this impulse of pure activity is the source and centre of all creation in the universe.

To Bergson, our mind, in its basic nature, is a tension of unified experiences and it unrolls from itself a series of images in an order of succession. If we analyse matter and other entities that exist outside our mind, we find them to be groups of ideas interconnected in various ways. The colour, shape, size and even the weight of matter are ultimately convertible to some ideas of our mind. The non-material objects, such as life and mind of other persons, are also some ideas only of our mind, or at least they conform to our ideas of life, mind etc. Whatever may be the real nature of entities outside our mind, they must be convertible to ideas so that they are known to us. Schopenhauer was right when he said, "No truth therefore is more certain, more independent of all others, and less in need of proof than this, that all, that exists for knowledge and therefore this whole world, is only object in relation to subject, perception of a perceiver, in a word, idea.... All that in any way belongs, or can belong to the world is inevitably thus conditioned through the subject, and exists only for the subject. The world is idea".[21]

If the entities that appear to be outside our mind were ideas only and nothing else they should have been like other entities, purely mental. But unlike purely mental entities, they appear to occupy portions of space. Though for us space is also an idea it implies its existence outside the mind. Any idea incorporated with the idea of space therefore implies that the entity so formed will be outside the mind. Thus we find that the superficial aspect of our mind, such as sensations, thoughts, emotions, etc. appear within the mind in an order of succession and ideas incorporated with the idea of space, such as the ideas of matter, etc. appear to be outside the mind in an order of succession. All these ideas also imply that they have a mind or principle of apprehension at the background which owns them.

The ideas of matter, etc. which refer to entities outside our mind do not depend on us in the sense that we cannot create such ideas at our sweet will. As these ideas are governed by some inexorable laws that are beyond our control. We may assume that they are a series of self-existing images. The external world, i.e. the space in which matter, etc. appear in time as an order of succession is thus reduced to ideas of the mind but these ideas are self-existing images independent of the mind. Are these images, then, within the mind or outside it?

As regards time, Bergson states categorically that it, as an order of succession, has its roots in the mind. According to him, the states in the mind may appear in an order of succession but objects in the external world are all present and present together. They are perceived as simultaneous and as extended in space, outside one another. The external world is the simultaneous togetherness of many objects existing side by side. The so-called successive states of the external objects are nothing but different simultaneous presentations of these objects arranged successively by the mind. The external world is thus devoid of time as an order of succession. The mind projects different elements of its tension in an order of succession for its closer attention to them for success in practical life. The ideas or images of time as an order of succession are thus within the mind. As the external world appears to be in time as an order of succession, time of such a nature also appears to be outside the mind. As time as an order of succession also is governed by certain inexorable laws, its images are self-existing in nature. Time as an order of succession is thus a self-existing image within the mind.

Time as an order of succession in the external world also implies the existence of space. In a sense, time as an order of succession is the root of our idea of space. Time, of whatever character it may be, is the condition for movement. Our idea of movement is impossible without the idea of time as an antecedent condition. Time makes movement possible. In the deeper aspect of mind, i.e. in the aspect of mind as a tension of various elements, also there is time but not as an order of succession. The elements, that are in movement here are not isolated from each other. They are intermingled and they interpenetrate each other. The time of such tension in the mind does not therefore mean an extended series of experiences coming one after another. As the elements of such tension are in the area of mind and as no separate rooms are required to accommodate them, the idea of space is absent here. Space means a receptacle to provide separate areas for the elements in question so that they may not penetrate each other. Interpenetration of elements does not necessitate the provision of separate rooms for the elements and as such no question of the idea of space can arise. When the intermingled and interpenetrated elements of the tension of mind are separated one from another, the said elements need separate areas for them. Thus when we relax the tension of our mind and separate its elements in an order of succession, we create thereby space to provide separate areas for the separated elements. Space also is thus rooted in the mind. But as the idea of space is governed by certain inexorable laws, it also must be a self-existing image within the mind. And time as an order of succession is the creator of space.

Now, the extra-mental objects appear to be in space and in time as an order of succession. But these objects, as we have seen, are nothing but a series of self-existing images. Can these images be within the mind? Do the objects of the external world lie outside the mind and create ideas in the mind by coming in contact with it through the sense-organs? If the external world lies outside our mind then we shall have to believe that the objects which are outside the mind occupy space which is within the mind. This is impossible and as such this position cannot be accepted. Again, the objects that appear to be in space are there because they are projected that way. The said objects were originally present in the tension of our mind as its various elements. When we relaxed that tension for closer attention to its elements, we allowed these elements to get isolated one from another, created space and time as an order of succession for the purpose and thus brought in our conception of externality. The objects of the external world thus appear to be outside our mind but actually they are within the mind.

A point in this connection draws our attention here. The external world, as it appears to us, comes in contact with us through our physical sense-organs. A physical sense-organ is an object of the external world. It appears to receive sense-data from the external world and it passes them on to our mind or the principle of apprehension. A man who is blind can have no image of light because his defective eyes are not able to collect the sense-data of light from the external world. Does it not mean that the sense-data come from outside and enter into our mind through the sense-organs? Are not the sense-data then outside the mind? It appears *prima facie* to be so. But a deeper analysis will show that the case really is quite otherwise. If we analyse a case of apprehension we shall have to admit that apprehension of sense-data depends not on our sense-organs but on our sense. A dead man does not apprehend a sense-datum even though his sense-organ may be in a perfect condition, whereas a man having intuitive power can apprehend a sense-datum even without the medium of a sense-organ. It shows that in an act of apprehension our sense is the real organ in use and not the sense-organ. Moreover, though the sense-organs and sense-data appear to be objects outside the mind, they are not really so because they occupy space which, according to us, is within the mind. An act of apprehension, according to us, depends on the faculty of exercising a sense. A man having no faculty for mathematics fails to understand a simple mathematical calculation and a man having no musical faculty cannot appreciate a piece of melodious music. Similarly, a man having no faculty of sensing cannot collect sense-data. The exercise of a physical sense-organ is only a condition to be fulfilled for sensing but it can be dispensed with by persons who have the intuitive power. In the cases of mathematical or musical faculties also certain conditions need to be fulfilled so that these faculties may be exercised. The physical sense-organs are not therefore the all-important media for us to come in contact with the external world. The role of sense-organs in the act of sensing in the ordinary way does not therefore contradict our view that the physical objects including the sense-organs are not without but within the mind. And our view that the objects which appear to exist outside the mind are actually within the mind does not also contradict the Bergsonian view that matter is an aggregate of self-existing images and it is precisely that which it appears to be. As images, the physical objects are within the mind and as images of a self-existing nature these objects are governed by certain inexorable laws and are not within the control of individual will. We thus reduce the whole universe into mind and its ideas of time, space and objects.

There are two phases of our mind, each containing ideas in some form or other. The first phase is a tension of various elements which, according to Bergson, is the deeper aspect of our mind. The mind in this aspect contains ideas but these ideas are not separated one from another. The ideas here are intermingled and they interpenetrate each other. Time is present here, though not as an order of succession but as duration. Space also is present there but in the potential form and the so-called physical objects are present in the form of interpenetrated ideas. When the tension of the mind is relaxed we enter into its second phase. In this phase of the mind we find sensations, emotions, thoughts, etc. which are separated, isolated, named and arranged one after another. Time here appears as an order of succession and space becomes an actuality. The ideas in the tension of mind now appear as images and we call them objects in space.

Now, the first phase or deeper aspect of mind cannot move on to its second phase as a matter of course. This point will be clear if we anlayse the view of Bergson that in every intense experience such as a violent love, a deep melancholy, a burning passion, there is a feeling of our entire existence being stirred and our whole being is moved in a way too deep for expression. It shows that sometimes from the second phase or superficial aspect of mind we pass on to its first phase. It also shows that when we are in the first phase of mind we fail to distinguish its various elements. But is it a universal truth? We do not think so, because, though most of us fail to distinguish the various elements of such a tension of our mind or to express our intense feelings, there are some who can do it. The ecstatic feelings of a poet or a musician are not less intense than the intense feelings of a man in violent love or in a deep melancholy. But whereas an ordinary person cannot distinguish the various elements of his deep feelings, a poet or a musician can. The journey of mind from its first to the second phase thus depends on the person concerned, which means that it depends on his desire and effort. A person can move on from the first to the second phase of his mind only when he so desires and makes efforts accordingly. The movement of mind from its second to the first phase also reveals the same fact. The love that transports a person from the second to the first phase of his mind is made violent by the person himself. The events that turn into a melancholic disposition depends on the mental make-up of the person concerned. The landscape that transports immediately a poet to the first phase of his mind may not have any effect on a person who has no aesthetic sense. The tune of music that can stir the mind of a musician may have no effect on a person who has no musical sense. It all shows that the movement of mind from its second to the first phase also depends on the person concerned, or, rather on his desires and efforts.

The desire and effort to do something is nothing but an exercise of one's will. A person passes on from the first phase of tension to the second phase of relaxation of his mind because he wills to do so for practical purposes. Everyone wants to live his life successfully and so directs attention to his environment in order to act properly. The will of a person thus works behind the movement of his mind from its first to its second phase. In the case of a violent love or a deep melancholy a person has the desire to enter into the depth of his mind and life and to study the situation from that broader perspective. Here also the will of a person works behind the movement of his mind from its second to the first phase.

We are now in a position to state in a general way that there lies will in between the two phases of mind, deeper and superficial. In its first or deeper phase the mind is a tension of various elements. We cannot isolate these elements, one from another. We cannot classify them nor can we assign familiar names to them. Is this stage not a stage of feeling only? It is not 'will' because we have placed 'will' at a later stage of the mind, where it acts as the motive force for relaxation of the tension of mind. This stage of mind is also not a stage of thought proper because thinking is a process of handling ideas, whereas the various elements in the tension of mind are yet to be ideas. If this tension of mind is neither 'will' nor 'thought' it cannot be anything other than a 'feeling' because thinking, feeling and willing are the only three constituents of our mind.

The deeper aspect of our mind is a state of 'feeling' which moves on through its aspect as 'will' to the superficial aspect of 'thought' dealing with ideas and images. These three stages of mind—feeling, will and thought—are not lineal in movement. This movement is circular in the sense that the last stage of thought creates again a fresh tension of feelings to be added to the original mass of feelings from where the whole process started. It starts from a mass of feelings and is ended in the creation of some more elements of feelings. This cycle of our mental life may be shown in the following manner:

Feeling ⟶ Will ⟶ Ideas, Thoughts etc. ⟶ Feeling

The consecutive order of the three aspects of our mind, as is shown in the above cycle, is true in view of the fact that a feeling cannot generate ideas, thoughts, etc. directly, nor can a will generate a feeling directly. A feeling generates ideas, thoughts, etc. through an exercise of the will and a will generates feeling only after the creation of ideas, thoughts, etc. It is, of course, a fact that the three aspects of our mind—feeling, will and thought—are actually intermingled in a concrete mind. These three aspects can be isolated one from another, but only theoretically.

Is feeling the starting point of our cycle of life? According to us, it is. The ideas, thoughts, etc. which appear in the superficial aspect of our mind has a tension as their background. To be more precise, it may be said that the said tension is analysed into ideas, thoughts, etc. and the motive force behind such analysis is the will. Can will be the starting point of our cycle of life? No, according to our analysis. Even Schopenhauer, to whom the universe is a manifestation of will, states that the basis of all willing is need, deficiency and pain.[22] Thus in Schopenhauer also the starting point of our cycle of life is not will but a feeling of pain. To Bergson, a tension of intermingled elements is the starting point of our cycle of life, which tension, as we have shown above, is not anything other than a feeling.

Now, a feeling can either be agreeable or disagreeable to us. Can it be anything else? No, because there are no other alternatives here. But what about an agreeable feeling that turns disagreeable to us or a disagreeable feeling that is turned into a feeling agreeable to us? When the agreeableness of a feeling is lessened moment after moment, a point of time will come when the feeling is no longer agreeable to us nor yet disagreeable. As at this moment we are in a state when we do not react to the feeling at all, will it mean that we have ceased to live? No, because we still live to react to the feeling the next moment, to feel the state as agreeable to us. What does then remain in us at that moment? It is not a feeling, because we do not react to it, nor can it be a will or a thought, because they come later. It can, according to us, be only a state of consciousness, pure and simple. It is a consciousness that was tinged with a feeling of agreeableness a moment earlier and will be tinged with a feeling of disagreeableness a moment later. The said consciousness is an entity that exists prior to the oncoming of any feeling, agreeable or disagreeable, to it. It is something that becomes either an agreeable or a disagreeable feeling to us later. We thus reduce our inner being into a consciousness, pure and simple. It is pure, because, it is not conscious of anything other than itself. It is simple, because, it has not yet manifested its products.

From the human point of view, which practically is the only way for us to look at things, we have analysed our conception of mind, matter, space and time and arrive now at the conclusion that the principle that lies at the root of this universe, we are in, is a pure consciousness. This consciousness, in course of its evolution, creates the cycle of our life. It is the principle that produces our mind and, through its activities, the so-called products of Nature—matter, space, etc.—in an order of succession.

References

1. Henri Bergson, *Time and Free Will,* Authorised translation by F.L. Pogson, M.A. (George Allen and Unwin Ltd., London, 1959), p. 231.
2. Henri Bergson, *Introduction to Metaphysics,* Translated by Mabelle L. Andison, Philosophical Library, Inc., New York, 1961, pp. 9-12.
3. Bergson, *op. cit.*, p. 100.
4. *Ibid.*, pp. 128-129.
5. *Ibid.*, pp. 226-227.
6. *Ibid.*, pp. 98-99.
7. *Ibid.*, pp. 120-121.
8. *Ibid.*, p. 125.
9. *Introduction to Metaphysics,* pp. 59-61.
10. *Time and Free Will,* p. 130.
11. *Ibid.*, pp. 132-133.
12. *Ibid.*, pp. 231-232.

13. Henri Bergson, *Matter and Memory*, Authorised Translation by N.M. Paul and W. Scott Palmer, George Allen and Unwin Ltd., London, 8th Impression, 1970, Introduction, pp. xi to xiii.

14. *Ibid*., p. 280.

15. Henri Bergson, *Creative Evolution*, Translated by Arthur Mitchell, The Modern Library, New York, 1944, pp. 220-221.

16. *Ibid*., pp. 259-260.

17. *Ibid*., p. 267.

18. *Ibid*., pp. 273-274.

19. *Ibid*., p. 291.

20. *Ibid*., pp. 284-285.

21. Irwin Edman (ed). *The Philosophy of Schopenhauer*, The Modern Library, New York, 1928, pp. 3-4.

22. *Ibid*., p. 352.

Chapter 2
Concepts of Time and Causality

That consciousness exists at the root of our universe has been admitted by Bergson. Thus he writes, "What, then, is the principle that has only to let go its tension—may we say to detend in order to extend, the interruption of the cause here being equivalent to reversal of the effect? For want of a better word we have called it consciousness. But we do not mean the narrowed consciousness that functions in each of us. Our own consciousness is the consciousness of a certain living being, placed in a certain area of space; and though it does indeed move in the same direction as its principle, it is continually drawn the opposite way, obliged, though it goes forward, to look behind".[1] The principle that lies and works behind the tension of feelings in our mind is a general consciousness and our individual consciousness is a part of it. This view of Bergson is in agreement with our conclusion drawn in the last chapter.

We may now see what Bergson does actually mean by 'consciousness'. As regards consciousness of a living being Bergson writes, "If we examine this point more closely, we shall find that consciousness is the light that plays around the zone of possible action or potential activity which surrounds the action really performed by the living being. It signifies hesitation or choice. Where many equally possible actions are indicated without there being any real action (as in a deliberation that has not come to an end), consciousness is intense. Where the action performed is the only action possible (as in activity of the somnambulistic or more generally automatic kind) consciousness is reduced to nothing".[2] And again, "Throughout the whole extent of the animal kingdom, we have said, consciousness seems proportionate to the living being's power of choice. It lights up the zone of potentialities that surrounds the act. It fills the interval between what is done and what might be done".[3] It appears from the above that consciousness is an instrument of action as also it is a manifestation of action. But Bergson himself reverses the position when he writes, "In this second hypothesis, as in the first, consciousness is still the instrument of action; but it is even more true to say that action is the instrument of consciousness; for the complicating of action with action, and the opposing of action to action, are for the imprisoned consciousness the only possible means to set itself free".[4] To define consciousness in a more specific way, Bergson writes, "The consciousness of a living being, as we have tried to prove elsewhere, is inseparable from its brain in the sense in which a sharp knife is inseparable from its edge: the brain is the sharp edge by which consciousness cuts into the compact tissue of events, but the brain is no more co-extensive with consciousness than the edge is with the knife".[5] "Radical therefore, also, is the difference between animal consciousness, even the most intelligent, and human consciousness. For consciousness corresponds exactly to the living being's power of choice, it is co-extensive with the fringe of possible action that surrounds the real action. Consciousness is synonymous with invention and with freedom".[6] "Consciousness, in man, is pre-eminently intellect. It might have been, it ought, so it seems, to have been also intuition. Intuition and intellect represent two opposite directions of the work of consciousness: intuition goes in the very direction of life, intellect goes in the inverse direction, and thus finds itself naturally in accordance with the movement of matter. A complete and perfect humanity would be that in which these two forms of conscious activity should attain their full development".[7] "Finally, consciousness is essentially free; it is freedom itself; but cannot pass through matter without settling on it, without adapting itself to it: this adaptation is what we call intellectuality; and the intellect, turning itself back toward active, that is to say free, consciousness naturally makes it enter into the conceptual forms into which it is accustomed to see matter fit. It will therefore always perceive freedom in the form of necessity; it will always neglect the part of novelty or of creation inherent in the free act; it will always substitute for action itself an imitation artificial, approximative, obtained by compounding the old with the old and the same with the same".[8]

The above statements show that consciousness, according to Bergson, is constituted of the activities of a living being, more particularly of its potential form. It is a living being's power of choice between two or more courses of possible action. This power of choice is exercised by the living being through its brain. Consciousness is synonymous with invention and freedom. Consciousness in man is pre-eminently intellect and as such it enters into the conceptual forms into which it is accustomed to see matter fit. Thus our intellect always substitutes for action itself an imitation artificial, approximative and obtained by compounding the old with the old. This intellectuality of our mind is due to the adaptation of our consciousness to matter. Intuition, like intellect, is another instrument of action for consciousness. Intuition is just opposite to intellect and it goes in the very direction of life.

Now, if consciousness of a living being is his power of choice between different courses of possible action then he will be conscious only when he exercises this power, i.e. only in acts that follow his intellectual deliberations and not in acts that are performed instinctively. But obviously it is not so. If to be living means to have consciousness a person must be conscious even when he acts instinctively. The lower animals and plants do not exercise any power of choice but then, do they not live? So, according to us, the statement of Bergson that in acts of the automatic kind a person's consciousness is reduced to nothing is only a figurative one.

The actual fact will therefore be that in acts where there is an exercise of the power of choice, we are conscious about our consciousness whereas in acts that are performed instinctively we are not so.

Now let us see how the consciousness of a person is connected with his life. According to Bergson, life in an organism is a continually growing action. He writes, "Regarded in what constitutes its true essence, namely, as a transition from species to species, life is a continually growing action".[9] And again, "Now the life manifested by an organism is, in our view, a certain effort to obtain certain things from the material world".[10]

Life is the effort of an organism to obtain certain things from the material world. Now, an effort to obtain certain things originates in a feeling of want because an effort to obtain things is always preceded by a desire to have the thing which again is preceded by a feeling of want. Life, then, is rooted on this feeling of want. A feeling, of whatever kind it may be, is a manifestation of consciousness. Consciousness thus is the root of life and its activities, both deliberative and instinctive.

Consciousness manifests a feeling of want. This feeling works as a principle behind life and sustains it through its efforts to obtain certain things. An action performed by a living being after deliberation makes him conscious about his life's activities. But instinctive activities are not preceded by deliberations and so they do not make the living beings so conscious. But consciousness is still there in him because it appears again when the living being becomes conscious of the work done, though instinctively. Thus, consciousness, explicit or implicit, is always present in a living being.

Though the consciousness of a living being has certain limitations in it because of its being placed in a particular area of space and in a particular point of time, it is consciousness all the same. The principle that works behind the detention of feelings in a living being cannot but be the same principle that works behind the movement of the universe as such, because a living being is a part and parcel of this universe. What is true for a part of the universe cannot be totally false for the whole of the universe. The principle that works behind the universe must be either identical with or similar to the principle that works in a living being. Bergson is therefore right when he says that the principle that works behind the process of detention is consciousness of which a narrowed form is present in a living being. To avoid confusion between consciousness and its narrowed form we may term them as prime-consciousness and consciousness respectively. Prime-consciousness is the first principle and our universe and consciousness come out of it in course of evolution.

The world a living being comes to confront is thus originated in his consciousness. As a living being has limitations, the consciousness in him is also a limited form of it. To reach unlimited consciousness, which, according to us, is the prime-consciousness and the first principle behind our universe, a living being is required to remove the said limitations. Bergson also arrives at this position when he says, "If our analysis is correct, it is consciousness, or rather supra-consciousness, that is at the origin of life. Consciousness, or supra-consciousness, is the name of the rocket whose extinguished fragments fall back as matter; consciousness, again, is the name for that which subsists of the rocket itself, passing through the fragments and lighting them up into organisms. But this consciousness which is a need of creation is made manifest to itself only where creation is possible. It lies dormant when life is condemned to automatism; it wakens as soon as the possibility of a choice is restored".[11] Here in these lines Bergson admits that there is consciousness even in those organisms who have no power of choice. Consciousness lies dormant in those organisms who are condemned to automatism.

Being the first principle behind this universe, prime consciousness contains everything that is experienced by a living being. Prime-consciousness is the cause for everything that appears as effect later. It is the creator of matter, life and the narrowed consciousness of living beings. But even before creating the evolutes, prime-consciousness has two principles involved in it. These are time and the principle of causality. According to Schopenhauer, time and space are principium individuationis. He writes, "In this last regard I shall call time and space the principium individudationis, borrowing an expression from the old schoolmen, and I beg to draw attention to this, once for all. For it is only through the medium of time and space that what is one and the same, both according to its nature and to its concept, yet appears as different, as a multiplicity of co-existent and successive phenomena. Thus time and space are principium individuationis, the subject of so many subtleties and disputes among the schoolmen".[12] But according to us, time is not principium individuationis. Prime-consciousness is or exists even before its creation of matter, life, mind, etc. To be or to exist is to be or to exist in time. The very notion of existence involves time in it. Nothing can exist without being in time. Prime-consciousness exists and so it must be in time. Do these two, prime-consciousness and time, then mean two different principles only attached to each other? Are they not reducible to each other? Our answer here is that although prime-consciousness and time appear to be two different principles they are not actually so. Time is only a part of the connotation of consciousness. Time means time-consciousness only. Consciousness is an all-inclusive notion and it means all that we may be conscious of. Taken in this sense of the term, prime-consciousness includes all. An individual consciousness does not include all because it has certain limitations in it due to its confinement in a particular area of space and a particular point of time. Prime-consciousness, in this sense, has no limitations in it. It does not cover a particular period of time nor does it occupy a particular area of space because time, space, and all come out of it. Time involved in prime-consciousness, cannot therefore be a parallel principle to it. Time stands subsumed in prime-consciousness and in this sense it is reducible to the latter. Can prime-consciousness be reduced to time? No, because time only means time-consciousness. Time is only one of the contents of prime-consciousness. Can either of these two principles be prior to the other? Can time be a principle that exists prior to prime-consciousness or prime-consciousness a principle that exists prior to time? Time is a principle that cannot exist prior to prime-consciousness because time is only a content of consciousness. Time means nothing without a consciousness of it. On the other side, prime-consciousness also cannot exist without being in time. The very existence of prime-consciousness involves time in it, because to exist means to exist in time. Though time is a content of prime-consciousness, it has the special status of being involved in the very existence of the latter. Time is not an evolute of prime-consciousness, like other evolutes that appear later.

Time, again, is a principle that is responsible for all creation and development. The very notion of time is that it flows. It is an impelling force that generates movement. It brings in movement in prime-consciousness which in turn generates or rather manifests its contents. The universe of matter, mind, etc. comes out of prime-consciousness because of the presence of time in it. Time, in this sense, is the cause of evolution of the universe from out of prime-consciousness. When the prime-consciousness of one moment passes on to the next moment, and which it must, prime-consciousness of the first moment, is changed. It is changed because a second moment is now added to it, which means the addition of a quantum of movement to it. Prime-consciousness of the first moment is thus not the same as the prime-consciousness of the second moment. The presence of time in prime-consciousness thus makes it a dynamic entity. We may, in this sense, say that prime-consciousness has the principle of movement inherent in it. Time is the impelling force that pushes prime-consciousness towards evolution.

Now, when prime-consciousness of the first moment moves on to the second moment, can its entity of the first moment be lost, never to be found again? No. Nothing can be lost because prime-consciousness has no 'outside' of it. Prime-consciousness covers and includes all and so there can be nothing outside it. All that is, is within prime-consciousness. In the second moment prime-consciousness has only a quantum of movement and change added to it. Prime-consciousness of the second moment retains everything that it had in the first moment and adds only a quantum of change to it. Prime-consciousness moves on in this way changing itself. A question may here arise: do only some identical moments added to prime-consciousness effect some real change in it? According to us, they do. If a thing does not remain identical all through its existence, we must say that it has undergone change. As prime-consciousness of the first moment moves on to the second moment adding thereby an extra moment to it, it is not surely the same as it was before. It is now changed. As time is a constituent part of prime-consciousness, the latter has the principle of change in itself. Prime-consciousness can never remain static for a moment. Can time be anything other than the principle of change? No, because the very concept of time includes change in it. If we go on dividing a moment of time into its parts shall we ever reach at a quantum of time which is static? No, because the smallest unit of time is also time and is in time. Time is thus change itself and/its presence in prime-consciousness makes the latter a dynamic principle.

Now, according to our hypothesis, prime-consciousness goes on adding quanta of time or change to it and so, in a sense, it is enlarging itself. Does this enlargement mean that there is a necessity of additional receptacle for the accommodation of additional elements? There would have been a necessity for additional receptacle if the enlargement had meant an addition of something physical. Also in that case there would have been the provision of a receptacle beyond the extent of prime-consciousness. But according to our hypothesis there is nothing beyond prime-consciousness. The change involved in the enlargement of prime-consciousness in question cannot therefore be of a physical nature. The change involved then must be a change of form only which requires no additional receptacle for its accommodation. So space, being the receptacle for physical entities, need not be manifested by prime-consciousness at this stage. Of course, space in an implicit form is present there in prime-consciousness, or it could not have evolved later. But time is present there in prime-consciousness in the explicit form to make prime-consciousness an existent entity.

Can prime-consciousness and time exist in an implicit form? No, because in that case they would require a receptacle to be accommodated in and the existence of that receptacle would imply explicit time in it. Time thus is existence itself. It implies existence as also movement or change. The existence of anything, explicit or implicit, involves the existence of time. Even the existence of time, in either explicit or implicit form, involves time in an explicit form. Time thus cannot have an implicit form. Prime-consciousness also can have no implicit form because its very existence implies that it is explicit. There is no substratum for it to be implicit in. Moreover, it is a fact that the very existence of prime-consciousness implies the presence of time in it and time, according to us, is always explicit. If time is time-consciousness only, then the explicit nature of time makes also prime-consciousness an explicit entity. Lastly, if prime-consciousness is implicit, then time, which is not anything beyond prime-consciousness, must also be implicit. Without being explicit, prime-consciousness cannot contain explicit time.

Now, we are justified in our earlier statement that space, matter, mind, etc. which appear in course of evolution later, pre-exist in prime-consciousness in an implicit form. Do their existence also not imply that they are ever-explicit? We are justified to consider space, matter, mind, etc. as entities present implicitly in prime-consciousness, because this form of existence is possible for them in view of the fact that they have a receptacle and substratum to be implicit in. Their very existence does not mean that they must be explicit. There is no contradiction in their existence in an implicit form. Space, matter, mind, etc. may exist in prime-consciousness in an implicit form even though the latter is explicit because the former are not the substratum for the latter. A seed which is explicit and visible may contain the life of a plant in an implicit form. To prime-consciousness, of course, nothing can be implicit because it owns and knows everything. But unless there are reasons for prime-consciousness to manifest them, they will only remain unmanifested.

The addition of new elements to prime-consciousness due to the presence of time in it means that it undergoes change every moment and produces some forms that are new. This novelty of form in prime-consciousness is, according to us, the change in form of the entities present there. The entities that undergo change and assume a novel form must have been present in prime-consciousness or they could not have appeared later from nowhere.

The evolutes of prime-consciousness must exist in it even before their manifestation. Does it then mean that the courses of evolution are not unknown to prime-consciousness? In one sense, it means so. As there is nothing beyond prime-consciousness, as whatever happens, happens only within its fold, even the knowledge of everything must be in prime-consciousness. But in another sense, it implies a self-contradiction. How can there be knowledge of the courses of evolution in prime-consciousness even before the actualization of that course? The knowledge of a thing cannot be full and perfect unless the thing is actual. If manifestation does mean actualization, then the unmanifested evolutes and the courses of their evolution are yet to be actual and as such cannot be known fully and perfectly. Prime-consciousness cannot therefore know fully the courses of evolution of the evolutes before their actual evolution. This anomalous position may be set right if we assume that the courses of evolution and the evolutes exist in prime-consciousness as only a possibility waiting there for becoming an actuality. Time here plays the role of an agent and pushes this possibility on towards its actualization. Time, in this sense, is the creator of all. It is the principle of invention. In Bergson's words, "Time is invention or it is nothing at all".[13] Prime-consciousness knows the evolutes and the courses of evolution even before they are actualized, but that knowledge relates only to the possibility of their being actualized and not what they are, when actualized.

Being the first principle of this universe, prime-consciousness contains all that is in the universe, though not all in the manifested form. All evolutes, except time, exist in prime-consciousness, unmanifested. They are implicitly present in the first principle or are in a form of 'possibility' only. But time is present in the first principle even before the manifestation of other evolutes. The very existence of prime-consciousness involves time in it and as such time is never implicit in it.

Now, prime-consciousness undergoes change every moment due to the presence of time in it. A quantum of time adds itself to prime-consciousness every moment forcing it to discontinue its earlier state of existence and to attain a new state. Is this change then in prime-consciousness only a change due to the addition of homogeneous quanta of time? Prima facie, it appears to be so. But in that case the change will be meaningless because nothing new comes out of it. For prime-consciousness it will practically amount to no difference whether only one quantum of time has been added to it or more such quanta. One or more quanta of time adds practically nothing to prime-consciousness except time and this time makes the prime-consciousness to continue to exist only.

We have shown earlier that time cannot be implicit because in that case the implicit time will require explicit time for its very existence. But in another sense time is implicitly present in prime-consciousness because time comes out of it in an even flow. If prime-consciousness has no principle of time in it, time cannot appear from nowhere. Our conception of time has thus two contradictory principles involved in it. It cannot remain implicit and again it cannot but be implicit. Let us now see what does 'time' mean when it is not explicit. If time is not explicit then it must either be non-existent or implicit. Time cannot be non-existent because in that case it cannot appear from nowhere. It must then be implicit. But implicit time requires explicit time for its very existence. Before removing this anomaly from the situation we propose to analyse first our notion of time. Time, as we conceive it, is infinite. Time cannot be finite, having a beginning and an end because, in that case, nothing could have existed before the beginning of time and nothing will exist after the end of time. But our notion of time does not necessarily mean that along with the absence of time other things must be absent. When a moment of time is gone, we conceive of it as gone forever. But, according to our hypothesis, the first principle of this universe, i.e. prime-consciousness contains everything and there is nothing beyond it. Nothing can enter into it and nothing can go out of it. A thing cannot therefore be lost. It may only be transformed. Time cannot, therefore, be an entity like other entities such as matter, mind, etc. This peculiar nature of time gives it a special status in the realm of entities in prime-consciousness. Now, to us the only thing that can be lost forever is a particular state or order of existence. We therefore propose to equate time with a particular order of existence.

If time is an entity like other entities that come out of prime-consciousness, it must have been existing in prime-consciousness in an implicit form so that it could come out of it in a continuous flow. But we have shown earlier that time cannot exist in prime-consciousness in an implicit form. Implicit time always requires explicit time for its very existence. We thus have to presuppose explicit time to make the existence of implicit time possible. This means that explicit time is an entity prior in existence to implicit time. As explicit and implicit time are not two separate entities but only one entity in two different forms we shall have to say that time existed first in the explicit form and then assumed an implicit form. But explicit time is not more than a quantum of it and from this quantum infinite time cannot be derived, which we must do because time flows continuously and is infinite. Time thus cannot be considered as an entity like matter, mind, etc. Time was, time is and time will be, even though it has no source to come from. Even if there is no time as a distinct entity in prime-consciousness, the latter will still exhibit time, for it exists. Time, even without being a distinct entity, is necessary for the very existence of all entities. So, according to us, time cannot but be existence itself. A quantum of time will then mean only an order or a particular state of existence and continuous flow of time will mean continuity in existence. Prime-consciousness exists and what it means is that it continues to exist because there is no reason for it to discontinue its existence. If, however, there is any reason for it to discontinue to exist, there will then be change in it. Again, its continuous existence means that there are various points in its existence or there are various states of its existence. Otherwise, there will be no difference between a particular state of its existence and its continuous existence. The continuous existence of prime-consciousness means that it has various states of existence and these states are distinct one from another. One particular state of the existence of prime-consciousness is not the same as another state of its existence. Continuous existence of prime-consciousness thus means that there is a continuous change in it.

Again, the existence of prime-consciousness implies that it has a reason to exist. This also means that it is to exist for the fulfillment of a purpose. If then the existence of prime-consciousness means a particular order of its existence only, it remains static at that state and fulfils no purpose. For the fulfillment of any purpose, therefore, prime-consciousness must change itself. Prime-consciousness is thus a principle that implies change in it. It exists and continues to exist assuming different orders of existence in an order of succession. And time not only means an order of existence but also a change in that order.

If prime-consciousness at a particular state is symbolically represented as X, then this X, as per our analysis above, changes itself and assumes the form X_1 and then the forms X_2, X_3, etc. in a consecutive manner. The forms X, X_1, X_2, X_3, etc. cannot be reversed, because X is transformed to X_1 and X_1 to X_2 and not X_1 transformed to X or X_2 to X_1. As X is transformed to X_1 and X_1 to X_2 we may say that X has produced X_1 and X_1 has produced X_2 or that X is the cause with reference to its effect X_1 and X_1 is the cause with reference to its effect X_2. Thus the very existence of prime-consciousness implies the principle of causality working in it.

When X is transformed to X_1, the state or order of the existence of X is terminated. But this termination does not mean that X is lost forever or that it is annihilated. Nothing, according to our hypothesis, can ever be lost in this universe. Prime-consciousness, being the first principle of this universe, contains and sustains everything. There is nothing beyond it or outside it. When X, representing a particular order of the existence of prime-consciousness, is changed into the form X_1, X is not lost but only absorbed in X_1. X_1 contains X and a quantum of change in it, and X represents an entity that has been transformed to X_1. Existence of the particular order X and its transformation to X_1 are what we mean by time as duration and time as a succession. As existence implies change in it, time as duration implies time as succession. As a change in X actually means a change in the order of the existence of X, i.e. its transformation to X_1, the said change of time implies a causal principle working in it. Time and the principle of causality are thus mutually implicated. The very existence of prime-consciousness not only implies its existence but also implies change and the workings of the causal principle in it.

When X is transformed to X_1, the existence of X as X comes to an end but there is no break in continuity of their existence. The existence of X is continued upto the coming into being of X_1, for, if there is a gap between the two one cannot be transformed to the other. As in X_1, X is not present in the form of X we may say that the cause X, after producing the effect X_1 is absorbed in it. Cause and effect thus do not belong to the same order of existence. Cause and effect are thus distinguishable, though not separable. They are distinguishable, because there is difference between a cause and its effect. But they are not separable, because, after being absorbed in the effect the cause cannot regain its previous order of existence.

The principle of causality is implied in the very existence of prime-consciousness. It does not work till a quantum of change is undergone by prime-consciousness. Prime-consciousness exists and so contains in it time as duration. It undergoes change due to inner necessity and so time now appears as succession. Due to change in its order of existence, prime-consciousness also reveals the principle of causality. It may therefore be said that prime-consciousness, which contains all in an implicit form, undergoes change due to inner necessity and goes on making the implicit entities explicit. The first evolutes of prime-consciousness are time as an order of succession and the principle of causality. Prime-consciousness in time, as duration, works through time as succession and the principle of causality to make explicit what were implicit in it. The process of transformation of the implicit entities to their explicit form, according to us, is the course of evolution of the world. The element of time in the evolution of the world is the order of succession in which the evolutes make their appearance in the explicit form. Time-order actually means the change in order of the evolutes.

Now, the evolutes in different orders of their existence are what the prime-consciousness only contained but did not reveal earlier. Prime-consciousness had them but only in their implicit form. The circumstantial forces that work on the evolutes at a particular stage determine the stage of evolution that will come next. Every stage in evolution is thus the only stage that could necessarily follow its previous stage. Evolution means the actualization of a possibility, the transformation of entities from the implicit to their explicit form.

Every stage in evolution is the necessary outcome of its previous stage. Prime-consciousness exists because there is no principle in it to negate its existence. Its existence implies the presence of time in it as duration. Its existence implies also change in it, or time as succession and the principle of causality. This change in prime-consciousness means the manifestation of entities that were present in it in an implicit form, and this is the evolution of the world.

References

1. *Creative Evolution*, p. 259.
2. *Ibid.*, p. 159.
3. *Ibid.*, p. 197.
4. *Loc.cit.*, p. 197.
5. *Ibid.*, p. 286.
6. *Ibid.*, p. 287.
7. *Ibid.*, p. 291.
8. *Ibid.*, pp. 294-295.
9. *Ibid.*, p. 142.
10. *Ibid.*, p. 151.
11. *Ibid.*, p. 284.
12. *The Philosophy of Schopenhauer*, pp. 75-76.
13. *Creative Evolution*, p. 371.

Chapter 3
Concept of Space and its Evolution —Concept of Energy and the Evolution of Matter

We have said that prime-consciousness is the first principle behind our universe but have said nothing about a material principle that may sustain it, whereas, in the case of individual consciousness we find it sustained by a material principle. Our reason here is this: In the case of our first principle, being the first in order of existence it is the stuff behind everything, including a material principle, whereas in the case of individual consciousness it is otherwise. What we find in the case of individual consciousness is that matter evolves life and consciousness in it. Being one of the many principles contained in prime-consciousness, the principle of matter cannot be presupposed in conceiving prime-consciousness. The principle of matter cannot therefore be the sustainer of prime-consciousness. But it is otherwise in the case of individual consciousness. In the case of individual consciousness, a material principle is to be presupposed. Individual consciousness is not a cut piece made out of prime-consciousness. Individual consciousness, according to us, is only prime-consciousness limited by a principle of matter. A principle of matter is thus presupposed in our conception of individual consciousness. In fact, a principle of matter creates individual consciousness and as such it appears as the sustainer of such consciousness.

The very existence of prime-consciousness implies the presence of time or change and the principle of causality in it.

Due to change in prime-consciousness, the various elements implicit in it may come in contact with each other in different ways. Prime-consciousness may also undergo change by manifesting entity or entities implicit in it. To undergo change due to inner necessity, prime-consciousness may follow any one of the above courses or it may follow both. If the first course is followed by prime-consciousness then the implicit entities in it will go on coming in contact with each other in various ways. The process may continue forever, though there may be repetitions of states in prime-consciousness. If the second course is followed by prime-consciousness, then the entities, implicitly present in it, will be made manifest and the manifested entities will go on forming composite objects. This process may also continue forever. There is a possibility of repetition of states in this process also. The same conclusion follows if prime-consciousness does undergo change by keeping a part of its implicit elements implicit, the other part being allowed to manifest themselves.

Of the three possible courses for undergoing change in itself, the one to be adopted by prime-consciousness depends on its 'will'. For prime-consciousness, there seems to be no other reason except its 'will' to determine its course of action. If there is to be change in prime-consciousness then that change can be effected by any one of the courses mentioned above. Any preference shown by prime-consciousness for a particular course must be due to a particular end in view.

If prime-consciousness would have adopted the first course of action, there could not be objects in their explicit form. But it is obvious that there are such objects. It proves that prime-consciousness did not adopt this course of action. The second and the third courses amount practically to the same position, because, even if all the entities implicit in prime-consciousness are made explicit we cannot ascertain whether or not some are yet to be made so. For us, therefore, the point here is whether there is manifestation of entities at all or not. The universe, as reflected in our consciousness, impels us to believe that entities have been manifested. According to us, therefore, prime-consciousness has adopted the second or the third course of action and it has done so keeping some specific end in view. The course of the world is thus the materialization of a plan of prime-consciousness.

If prime-consciousness moves on in the line of manifesting the implicit entities contained in it, what will it manifest first? It cannot be individual consciousness because such consciousness as an explicit entity has in it a principle of limitation which has not yet been manifested. For the purpose of creating conscious beings, or individual consciousness, manifestation of matter is a necessity. Again, as consciousness is already a manifested entity, only matter can be the entity, to be manifested next. By the word 'matter' we do not here mean matter as is conceived in the sciences. What we mean by matter is any entity other than consciousness. It also means anything that can exist in consciousness in an implicit form.

Let us now determine the process through which an entity, implicit in prime-consciousness, goes in order to be explicit and how does an explicit entity differ from its implicit state. For an entity to remain implicit in prime-consciousness means that the entity is integrally one with other entities in prime-consciousness. It is so because the various entities in their implicit state cannot be isolated one from another. The entities form an integral mass in consciousness, because consciousness has a continuity in it. Consciousness cannot mean bits of consciousness joined together, for, in that case a connecting medium will be necessary which in turn will need another connecting medium and so on *ad infinitum*. This will lead only to an infinite regress. Having, therefore, continuity in it, consciousness must have its contents integrally connected with each other, because consciousness means consciousness of entities. The entities, implicit in prime-consciousness, form an integral mass and so no entity is distinct and well defined. But when the same entity is manifested and made explicit it assumes a distinct form. In their implicit form the various entities interpenetrate each other but in their explicit form they cannot do so.

In view of the above, we may say that an explicit entity requires an area exclusively for its own, an area which cannot be occupied by any other entity. As each of the explicit entities is distinct and separate from other entities, each such entity must, therefore, have a separate area of its own. The existence of space is thus presupposed in the existence of entities in their explicit form. Therefore, the first entity manifested by prime-consciousness (implying time, or change and causality in it) is, space.

Space is the container of entities other than consciousness. Does it then mean that consciousness lies beyond space? No. It only means that consciousness does not require space as its container. As a product of consciousness, space cannot contain consciousness. It is rather consciousness which contains space as one of its constituent elements. How from out of prime-consciousness does space actually come into existence? When prime-consciousness makes an implicit entity explicit, it isolates the entity from the integrated mass of other implicit entities. The isolated entity now assumes a separate and specific form of its own. A medium is necessary to accommodate this new form. Space, according to us, is that medium. The implicit entities exist in prime-consciousness which acts as their medium and container. The same entities, after attaining a new form due to manifestation, move on to the medium of space, which acts as their container. As the said entities do now lie outside the integrated mass of implicit elements they must be in a different medium or level of consciousness. This level of consciousness is our notion of space.

The evolution of space and the manifestation of matter, or entities which were in the implicit form, are simultaneous. If matter is not made explicit, there is no necessity of the evolution of space. There cannot, therefore, be matter without space or space without matter. Space and matter are correlated entities.

Can space, as a particular level of consciousness, be an ever-explicit entity like prime-consciousness, time and causality? It need not be so. Space comes into existence only when the implicit entities are manifested by prime-consciousness. Being correlated to matter in the explicit form, space cannot be explicit when matter continues to remain in its implicit form.

Space is not matter because it has no boundary. It is a boundless entity and cannot have an outside of it, whereas matter has a limited extent and boundary. Being thus a non-material entity, space cannot be anything other than consciousness. As the implicit entities are contained by prime-consciousness, so the explicit entities are contained by space. Now, prime-consciousness, time and causality are, according to us, ever-explicit entities. Does it then mean that space contains these entities also? No, because space is not more than a particular level of consciousness and as such cannot contain the whole matrix of prime-consciousness-time-causality. Space only stands as an entity between prime-consciousness and matter.

Let us now see whether space is basically matter or consciousness or both or neither. Space, according to us, cannot be matter because, unlike matter, it has no boundary and it is not impenetrable. An entity in its explicit form has a distinct and specific nature of its own. It has, therefore, a specific extension and a boundary and it is confined therein. It also means that beyond the extension of one entity, there may be other such entities having specific extensions of their own. The specific extension and boundary are exclusive properties of an entity. The entities may interchange their positions in space but one entity cannot occupy the space occupied by another, for, in that case the entities will have no exclusive extension of their own, and as such cannot be considered as explicit. For an entity, to be explicit and to have a specific extension is to be also impenetrable. But in the case of space it is otherwise. Space cannot have a boundary or limited extension, for, in that case, there must be space beyond the boundary of space and it is impossible. Again, space is a penetrable entity because it accommodates the presence of other entities in it. Being thus an entity basically different from matter, space cannot be material in nature. It may here be asked: if to be explicit, an entity is to have a specific extension and to be impenetrable, how can we say that space is an entity made explicit by prime-consciousness? Does it then mean that as an entity space is not like other implicit entities that may be manifested by prime-consciousness? We propose to give our answer to this question after determining the nature of space.

The striking features of space are that it is an unbounded entity and it has a pervading nature in its own sphere. It has been made to come into existence and to act as a substratum for explicit entities and it covers up all such entities. The explicit entities may be in space in many different collocations but space remains the same. Its nature is not lost in any sense. Space as an entity is penetrable. When an entity occupies a particular area of space, that area does not cease to be space. This is not so in the case of material entity. A particular area of a material entity cannot be occupied by another material entity keeping the first entity intact. A particular area of space cannot be occupied by two explicit entities simultaneously and it is due to the fact that the said entities have specific boundaries of their own. Had they been unbounded they could have occupied the same area of space simultaneously, for in that case they would have coincided exactly with the whole area of space.

For the perception of physical entities, including even my own body, I am required to put them in a spatial order. But am I to do the same while perceiving consciousness, both mine and that of others? I can come to know my own consciousness directly and without placing it in a spatial order. Through an elaborate inferential process I come to know the consciousness of others but I do not place such consciousness also in any spatial order. Spatial order is thus a prerequisite for the perception of physical entities only. But space, as we have shown earlier, is not a physical entity. It may then be a particular form of my consciousness only which is necessary for my being aware of physical entities. Moreover, not being a physical entity space cannot but be contained in consciousness.

Let us now see whether space is only a form in consciousness or it is a form of consciousness. I may be aware of my own consciousness and of a physical entity consecutively. The ways involved in my awareness on these two occasions are not the same. The main point of difference here is that on the former occasion I enter into my own self and on the latter I move away from it. When I focus my attention on a physical entity and become aware of it I do of course continue to be aware of my own consciousness but when I focus my attention on my own consciousness I do become aware of it only. The more I withdraw my attention from the physical entities the more intensified will be my awareness of my own consciousness. When the withdrawal of my consciousness from physical entities is complete the consciousness of my consciousness is also complete. Before the evolution of space and the physical entities consciousness is conscious of itself only. The oncoming of physical entities necessitates some changes in consciousness bringing therein some conditions so that the said entities may be perceived by it. Consciousness, so conditioned, is the consciousness of space. Space thus is a form of consciousness distinct from its original form. With the evolution of space necessary arrangements are made complete in consciousness for the oncoming of physical entities and their perception.

What now is the relation between prime-consciousness and the new form of consciousness, or space? Does the evolution of space mean that prime-consciousness itself is now transformed into space? According to us, it is not so. Being a form of consciousness, space is brought out from prime-consciousness and as such space is only an entity within its fold.

Being a form of consciousness, space is infinite in extent. The extent of prime-consciousness is not at all affected due to the evolution of space in it, though it undergoes thereby a change. Again, being only a modification of prime-consciousness, space is not outside it. The difference in the state of prime-consciousness before and after the evolution of space is that one of the elements contained in it was implicit before such evolution and explicit after it.

When prime-consciousness unfolds time, causality and space, it undergoes a change in it. Does this change in prime-consciousness mean that its previous state is lost forever and is absorbed in its later state? Can a state of prime-consciousness be lost when it attains a subsequent state due to change? Prima facie, it appears that the original state of prime-consciousness is lost when it attains a new state. But in that case, the said state of prime-consciousness requires something to be lost into. There being nothing besides prime-consciousness, such a loss of the original state of prime-consciousness is obviously impossible. The original state of prime-consciousness is thus not lost but is only absorbed in its subsequent state.

As we have said earlier, the evolution of space is a pre-requisite for the evolution of matter, or anything other than consciousness, and this evolution is simultaneous with the evolution of matter. Through the evolution of space, prime-consciousness brings in the principle of limitation in it. We have so far derived the principles of change, time, causality and space from prime-consciousness. Out of these principles, the first three are the implications of prime-consciousness itself. Prime-consciousness being an explicit entity, its implications, or change, time and causality are also manifested entities. But it is otherwise in the case of space. Being a form of consciousness, space was implicit in prime-consciousness and made manifest by latter for the manifestation of other implicit entities contained therein. The said entities are the contents of prime-consciousness and as such the latter is conscious of them. Being something prime-consciousness is conscious of, these entities are distinguishable from prime-consciousness, though not separable from it, because they are mutually implicated. Prime-consciousness without anything to be conscious of is a mere blank and objects to be conscious of without a principle of consciousness is a meaningless abstraction.

Now, what entities are implicit in prime-consciousness and are its contents? These entities must be something that lie hidden in prime-consciousness and constitute its contents. Without these entities prime-consciousness will only be an empty principle or a vague abstraction. In a sense the said entities are the very principle of sustenance for prime-consciousness. Now, we have shown that prime-consciousness implies the principles of change, time and causality. We have also shown that space is only a form of consciousness. Change, time, causality and space are thus elements of the very constitution of prime-consciousness and as such cannot be considered as contents of the latter. We are, therefore, to search for some other principle in prime-consciousness that exists in an implicit form and may truly be called the content of prime-consciousness. We find such a principle in energy which exists implicitly in prime-consciousness. Energy, as defined in Physics, is the capacity for doing work. Work, in this sense, is involved in the very existence of a thing. If no work is involved in the existence of a thing, then a transition from existence to non-existence of the thing and vice-versa will not involve a change in the process. If 'to exist' is not to do anything, then there will be no difference between 'to exist' and 'not to exist' because both will mean 'to do nothing'. The difference between existence of a thing and non-existence of the thing is that the former does exist or does some work and the latter does not exist or does not work. It may be argued that if to exist is to do something then a thing must first exist in order to do something. Our counter-argument here will be: the thing thus does exist or it does something. Existence and work are connected notions. Existence without work or work without existence is impossible. When a thing continues to exist it goes on existing or goes on doing something. If it is said that a thing that exists does something only when it ceases to exist, then we shall have to admit that to exist or to come into existence is also to do something.

In view of the above we may now say that the very existence of prime-consciousness implies work done by it and so, in terms of physics, it has energy. Prime-consciousness continues to exist, or continues to manifest energy. Again, prime-consciousness implies change in it. Change is the movement from one state to another and to move from one state to another is to do some work. Change in prime-consciousness thus implies the existence of energy in it.

Can prime-consciousness manifest its energy all at once? No, because it is impossible. If prime-consciousness manifests all its energy at a time it will not even be able to exist, for, to continue to exist is to move from one state to another for which energy is necessary. If all energy is spent at a time, then there will be no energy left in prime-consciousness for its movement from one state to another. Prime-consciousness will then have to cease to exist. But to cease to exist also some more energy will be necessary. There being no energy left in prime-consciousness, it will not then be able to cease to exist. Prime-consciousness thus can neither exist nor cease to exist. Prime-consciousness cannot therefore manifest its energy all at once.

Prime-consciousness manifests its energy gradually in course of its changes. There must be energy hidden in prime-consciousness or it could not have manifested the same. Though prime-consciousness is explicit, the energy contained in it is not all explicit, because it does not manifest energy all at once. The bulk of energy that remains hidden in prime-consciousness may be considered as implicit in form. The bulk of energy manifested by prime-consciousness due to change cannot be separated from prime-consciousness and its process of change. Like prime-consciousness, therefore, this energy has no limited extent or boundary and as such it needs no principle of limitation or boundary. But when prime-consciousness releases or manifests energy as a distinct entity it needs a principle of limitation or a boundary. This principle of limitation, according to us, is space.

Manifestation of energy as an entity distinct from prime-consciousness, though not separable from it, is not a necessary act for prime-consciousness. Prime-consciousness may continue to exist for ever even without manifesting energy as a distinct and bounded entity. In manifesting a bulk of energy as a distinct entity, therefore, prime-consciousness exercises its will deliberately. And it is not unreasonable to believe that manifestation of energy in this manner follows a plan designed by prime-consciousness. A will, according to us, is always preceded by a deliberation. A deliberation ends in a will only when out of many possible courses of action one particular course is chosen because of its being considered as the most suitable to fulfil a purpose. The process that determines one out of many possible courses of action is not simply an anthropomorphic explanation but is also logical. Whenever out of many possible courses of action one particular course becomes actual, there lies some reason for that. What can that reason be? It cannot but be that the course of action made actual was the most suitable or that it served a purpose and followed a law. Suitability of the course of action means that it is the most conducive to materialize a pre-planned design. If the action is not directed towards the attainment of a definite purpose, then it can neither be suitable nor unsuitable. Fulfilment of a purpose presupposes a will and a deliberation behind it. Now, can the course followed by 'an action' be only accidental in the sense that the action follows no plan and fulfils no purpose? No, because, even a so-called accident cannot be accidental in the strict sense of the term. There must be reasons behind an accidental action also. Every action is determined by its antecedent conditions and this determination follows a law. Every action thus follows a law, though it may not be known to us. An action is called accidental when it cannot be explained properly by pointing out the law that governs it.

Let us now see whether or not the law that governs an action implies that the action serves a purpose. If an action is governed by a law, it may be said that the action serves at least the purpose of being governed by the law. It has at least the purpose of being kept within the jurisdiction of a law. The action thus serves a purpose and as such the course of action is not the product of a chance. If it is said that an action follows no law, rather that it follows an arbitrary course determined by nothing, then our reply to it will be that the action then follows the law of keeping itself beyond the jurisdiction of any law. Thus in every case an action is governed by a law and it serves a purpose. Fulfilment of a purpose involved in any action implies a will behind it and also a deliberation. Manifestation of energy in a separate and bounded form is therefore willed by prime-consciousness and this action is designed to fulfil a purpose.

Now, prime-consciousness and its implications, namely, change-time-causality-space are all, obviously, entities of the nature of consciousness. Consciousness also implies contents of consciousness or objects. "Consciousness" is a noun abstracted from the verb "to be conscious of". "Consciousness" will therefore be a reality only when it has also contents in it. It may of course be said that consciousness may have only itself for its content. In other words, prime-consciousness may be of the nature of self-consciousness. But in that case an abstract entity will have only abstractness as its content and will not in any way or in any sense be a concrete reality. In order, therefore, to be a concrete reality prime-consciousness must have its content. Energy, according to us, is that content.

Prime-consciousness is not merely an abstract principle. It is a concrete entity having energy as its content. Prime-consciousness is conscious of energy.

When prime-consciousness makes explicit a part of its energy, the said energy is no longer an element of the integrated mass of various elements in prime-consciousness (by 'various elements' we mean the elements that are implicit in energy, lying in wait to be made manifest later). This explicit energy is now in a differentiated and specific form and as such has a boundary or limitation. This energy is also now existing at a separate level of consciousness, which is the consciousness of space. Space has therefore been evolved by prime-consciousness to accommodate this explicit energy.

When this energy is placed in space, or rather that the principle of limitation works on energy, not only it is that the bulk of energy confines itself within a boundary but also the constituent parts of it undergo a change in that way. Not only will this energy as a whole but also its constituent elements will now maintain a specific extent and nature of their own. The principle of limitation works not only on the bulk of energy as a whole but also on its constituent elements and as such this energy cannot now exist as an undifferentiated mass of elements. It now becomes an assembly of its differentiated elements. These differentiated elements are the invisible units of corpuscles of energy. These units are separated one from another and each of them has a specific extent, limit and capacity of its own.

The units of energy are governed not only by the principle of limitation but also by the laws or principles inherent in them, i.e. by the principles they imply by virtue of their being units of energy. The principle of limitation works on these units and makes them maintain their specific extent. And the principles inherent in a situation where various units of energy are present make the said units combine with each other in various ways. These latter principles are of course consequents or corollaries of the former. The principles of combination follow the principle of limitation as a matter of course.

The impelling force that makes the energy combine and part with each other comes from their very existence, for change, according to us, is the law of existence. The units or corpuscles of energy exist and undergo change to exist. This change brings in movement in them and as such they come in contact with each other and also sometimes recede away from each other. Permutations and combinations of the said units of energy thus take place and as a result their integration and disintegration give rise to the formation and breaking up of various chemical elements. By "chemical elements" we do not mean the elements of the Periodic Table. Chemical elements do not necessarily mean the elements like Hydrogen, Oxygen, Iron, Mercury, Gold, etc. which are comparatively stable forms of the corpuscles of energy in their higher and complex combinations. There are many stages the corpuscles of energy pass through till they reach these stable molecular forms. The formation of electrons, protons, etc. and the formation of atoms are such stages discovered by the scientists. But these are not necessarily the only stages the corpuscles of energy pass through to reach the molecular stage. By "Chemical elements" we therefore mean any or all of the materialized forms of energy, not necessarily the electrons, protons etc. or even the charges of electricity as specified by the scientists. There may be chemical elements, in our sense of the term, which are smaller than even the charges of electricity. Now, in course of further combination in various complex ways, the chemical elements form the atoms and molecules which are studied by the physicists and the chemists. The atoms and molecules are thus reservoirs of energy in a stable form.

It may here be asked why we bring in the question of an impelling force to make the corpuscles of energy come in contact with each other when they themselves imply work. Our answer to this question will be that they are only potential energy to be made kinetic by an impelling force. Moreover, energy is only a capacity for doing work, which does not necessarily mean that it is work actually done. An impelling force is therefore necessary to bring in movement in the corpuscles of energy. As we have stated earlier, this impelling force comes as a result of the workings of the principle of limitation on them.

Now, what about the laws that govern the corpuscles of energy and their various compounds? Are these laws independent or are covered by the laws of prime-consciousness? Our answer to this question lies in the fact that energy as a whole is not an entity independent of prime-consciousness. We have shown earlier that prime-consciousness implies energy in it. As an implication of prime-consciousness, energy is not therefore an independent and self-existent entity. The bulk of energy in question has been made explicit by prime-consciousness after deliberation and by an exercise of its will. In order to make the existence of energy in the explicit form possible, prime-consciousness has evolved space to accommodate such energy therein. The evolution of space, the evolution of energy as a spatial entity and the workings of the principles of limitation on it, etc. are all due to the will of the prime-consciousness. Energy, both implicit and explicit, are thus within the fold of prime-consciousness and not beyond it.

We have shown earlier that the existence of prime-consciousness implies the presence of energy in it. Let us now see whether or not energy also implies the presence of consciousness in it. Being the content of prime-consciousness, energy is reflected in it. Prime-consciousness is conscious of energy and it means that they are related to each other. If it is not so, how can one be reflected in the other or one be the content of the other? Now, when there is a relation between prime-consciousness and energy, a medium to relate them must be conceived. But the medium cannot be a third entity connecting prime-consciousness and energy, for in that case a fourth and fifth media are to be conceived to connect these fourth and fifth entities with the third and so on, and this situation will ultimately land at an infinite regress. The medium in question must therefore be conceived to have an element common to both prime-consciousness and energy so that it may have the capacity to relate the two entities. Not being a third entity, the medium must be present in both prime-consciousness and energy. The presence of a medium in energy thus means the presence of prime-consciousness in it. In fact, anything that may be reflected in prime-consciousness must have consciousness in it. Otherwise, the thing cannot stand in any relation to it. A thing, directly opposed to consciousness, cannot, according to us, stand in any relation to it. It may be said that "opposed to consciousness" is also related to consciousness because the former, being also an item of the latter, is within and not outside it. But in that case "opposed to consciousness" will not be opposed to consciousness. It will rather be an entity of the nature of consciousness.

Now, let us see whether an object may exist outside consciousness. According to us, it cannot, because in that case such an entity will neither be known nor unknown to us and no notion of it will be reflected in consciousness. If the entity in question is known to us it cannot be outside consciousness. If the entity is unknown to us, then also a notion of it is reflected in consciousness. Being then a known entity, the notion is inside and not outside consciousness. Now, for consciousness, an object is nothing but the notion of the object. A notion of the object being present in consciousness, the object cannot be said as unknown. For consciousness, therefore, an object outside it is an impossibility. And an object of which there is no notion reflected in consciousness is nothing but a non-entity to it.

Can consciousness be conscious of a thing unless the thing makes it so? Can we be conscious of a thing unless it makes us so? If we can, then it will be possible for us to make and unmake ideas of things according to our will because things, for us, are nothing but ideas. As the existence of objects does not depend on our will, as the objects have a recalcitrant nature, we are forced to believe that we can be conscious of a thing only when the thing makes us so. Of course, it also is a fact that all the conditions that make us conscious of an object are not determined by the object. Some of such conditions are determined by us in the sense that an individual's consciousness is to be made ready for it.

When I come in touch with an object and somehow become conscious of it, "something" of the object enters into my being. If the reflections of different objects in my consciousness are not different how do I distinguish one from another? Now, what can the "something" in question be? The reflection of the object in my consciousness is after all an idea of the object only. Though my idea of an object includes that it is outside my consciousness it cannot actually be so, because the object's existence outside is nothing but an idea to me. The idea being an element in my consciousness the object I deal with is within and not outside my consciousness. Whatever nature an object may have it can be reflected in my consciousness only in the form of an idea, i.e., in the form of an element of consciousness. Now, if the idea of an object is basically anti-consciousness in nature it cannot develop a relation with my consciousness. But the idea in question has actually developed a relation with my consciousness in the form of a container and contained, because my consciousness contains the idea of the object. If my consciousness is P, then, being anti-consciousness, the idea of the object will be not-P. And in no case can P contain not-P in it. The idea of an object cannot therefore be anti-consciousness. Even though the idea of an object is not the object itself, it must have something of the object, something that makes the idea a representation of the object. It is true that the idea of an object is not the object itself, for, in that case the object, having been reflected differently in different individual's consciousness, will have contradictory qualities in it, which is impossible. Again, if the idea of the object is the object itself, then the same object will be present in different places at the same time. But this is impossible in view of the fact that an object is a spatial entity having a specific boundary of its own and as such is confined in a particular area of space. The idea of the object is thus only a representation of the object only. But though only a representation of the object, the idea must have something in it, which it has derived from the object and which has some specific element of the object. What can this specific element be? It cannot be the energy the object is composed of, because, in that case the same energy will have to be present in the consciousness of different individuals at the same time and it is obviously impossible. The specific element of the object cannot also be its quality, evolved in the object due to a particular collocation of energy in it, because the quality is attached to the object and cannot leave it to enter into my consciousness. The only other thing present in the object is its specific form. This form is neither the energy contained in the object nor its qualities, though it is connected with both of them. This form must be a non-material entity because it enters into my consciousness as an idea and may also be present in the consciousness of other persons at the same time. Not being energy this form must be of the nature of consciousness because there is no entity in this universe other than energy and consciousness. If this form is considered as an entity other than energy and consciousness then it will require a medium to connect it with energy and consciousness, for it is present in both of them. Being an extra entity this medium again will require a medium to connect it with the entities concerned and so on *ad infinitum*. The form of the object in question is thus not an entity other than energy and consciousness which are mutually implicated. It is an inseparable element of energy contained in the object and it is also present in my consciousness. Being a non-material entity, as shown earlier, the specific form of the object is of the nature of consciousness, and being limited due to a specific character, it is of the nature of energy with spatial limitations. Again, the specific form of the object is neither pure energy as it is non-material nor pure consciousness as it is limited by its specific character. An object which is composed of energy and has a specific form thus contains elements of consciousness in it. Energy in a spatial form thus implies in it the presence of consciousness as limited by individuality.

In the case of infinite energy and its reflection in prime-consciousness, the position is not basically different, though the relation between them is direct. They are related to each other not through any specific form as in my case. If there is no element of consciousness in infinite energy, they cannot in any way be related to each other, because, there cannot be a relation between P and not-P. P cannot in any case contain not-P in it. Thus, in view of the fact that prime-consciousness implies the presence of energy in it, infinite energy also contains elements of consciousness in it.

Consciousness implies some contents it is conscious of and as such the contents are parts of consciousness. The contents of prime-consciousness being energy, the latter is a part of consciousness. In self-consciousness, consciousness is conscious not of the blank form but also of its contents. If the objects, or energy in them, are anti-consciousness by nature they cannot be parts of consciousness. Consciousness must therefore be present in energy or energy must imply consciousness in it.

One more point needs clarification before we conclude this chapter. When a bulk of energy is made manifest and accommodated in space does it sever all its connection with prime-consciousness? How is spatial energy, or the material world, connected with prime-consciousness and infinite energy? Space, which, according to us, is a particular form of consciousness, is a part of prime-consciousness all the same. Space is a constituent element of prime-consciousness and not anything separated from it. Prime-consciousness and space are thus integrally connected with each other. Similarly, explicit energy is energy all the same, though it is in a form different from that it was in prior to its becoming explicit. Energy in a spatial form is a part of infinite energy contained in prime-consciousness. Spatial energy or the material world is thus only a constituent element of prime-consciousness and its infinite energy. It is a particular form of a part of prime-consciousness and infinite energy and as such is contained in the latter. The material world cannot thus be disconnected in any way from its source, the prime-consciousness and its infinite energy.

We have shown that prime-consciousness implies energy and its content and energy implies the presence of consciousness in it. Does it mean that they together constitute only one entity? They do, according to us. As consciousness implies the presence of energy in it and as energy implies consciousness in it, pure consciousness and pure energy are only non-entities. Consciousness is a form of which energy is the content. One without the other is a meaningless abstraction. As each one of them implies the other, they together constitute the reality in them. They are distinguishable by virtue of their different functions but are not separable from each other.

Chapter 4
The Forces that Work on Matter—Evolution of Life and Mind

Four principles, in the form of forces, work on the corpuscles of energy which are in space. The first and foremost of these principles is the principle of change. Change, as we have shown earlier, is the law of existence. An entity that exists must undergo change to maintain its existence. Being some indivisible units, any change in the corpuscles of energy move about in space. And as a result of this movement in the corpuscles of energy space also undergoes some change. Space, which is a form of consciousness, is affected whenever there is any change in its contents. Consciousness and energy being mutually implicated entities, change in one implies and brings about a change in the other. Consciousness and energy together form a whole which is the reality of our physical world.

The second principle that works on the corpuscles of energy and space is the will of the Absolute or prime-consciousness and its infinite energy. A bulk from out of its infinite energy was manifested by the Absolute with an end in view and a particular level or form of consciousness was evolved to accommodate that energy. The inner workings of the manifested world in general and movements of the corpuscles of energy in particular are guided by the will of the Absolute. No movement in the corpuscles of energy is accidental in the sense that it serves no purpose. Every movement in the corpuscles of energy is oriented towards the fulfilment of a purpose planned by the Absolute. The movements of energy in a particular area of the physical world may seem to be disharmonious or even chaotic but when set in the total plan of the Absolute they are not so.

The third principle is a corollary of the second one. To conform to a harmonized scheme of the Absolute the corpuscles of energy must follow certain principles or laws evolved or formulated by the Absolute. These are the laws that the scientists discover. The bases of these laws are the principles of change, time, causality, space and energy. The principles of time and change are the very laws of existence. The corpuscles of energy exist and so they undergo change to exist. The said change implies the existence of time in the process. Every state of energy passes on to its next state and every state is determined by its previous state. The principle of causality is thus involved in the movement of energy corpuscles. Now, the corpuscles of energy are in space, or in a particular level of consciousness. The corpuscles of energy are energy in a manifested form and for their manifestation a particular level of consciousness, or space, was evolved. Space is therefore connected with the very existence of manifested energy. Moreover, the corpuscles of energy move about in space and so their movement, which means also their existence, implies the existence of space in the process. The corpuscles of energy are functionally different from the principles of consciousness. Being the contents of prime-consciousness the corpuscles of energy are not prime-consciousness itself and as such they are guided by certain principles specifically their own.

Out of all the principles that guide the corpuscles of energy in their movement, the fourth one is the most important and fundamental in nature. It is the principle of manifestation. This principle brings the corpuscles of energy into existence and sustains them in their manifested form. The other three principles that work on the corpuscles of energy are also under its sweep. The principles underlying the existence of space and the corpuscles of energy are an elaboration of the principle of manifestation and as such are only the outcome of it. The principle of change, time and causality, though not an outcome of the principle of manifestation, are also under its sweep in the sense that all their workings are oriented to keep the corpuscles of energy in their manifested form. Only those laws of the principles of change, time and causality which are relevant to the principle of manifestation work on the corpuscles of energy in space. Thus, the guiding factor in the workings of the laws of change, time, causality, space and energy on the corpuscles of energy in space is the principle of manifestation.

Now, the principle of manifestation, according to us, is the principle of separation of entities from each other. Manifestation means the confinement of a unit of the entity within a definite boundary. It is the imposition of a limitation on the entity manifested, the imposition of a boundary on them. The entity, before manifestation, was one with the integrated mass of other entities and so it was undifferentiated, unlimited and indefinite. After manifestation, the entity is definite and so limited within a boundary. The principle of manifestation makes an entity pass on from an indefinite homogeneity to a definite heterogeneity.

If we arrange the four forces, enumerated above, in a causal order, they will be: (1) change, being the very law of existence, (2) will of the Absolute, or prime-consciousness coupled with the infinite energy contained in it, (3) Principle of manifestation of energy, and (4) Scientific laws that operate on the mass of manifested energy. Though all these forces are equally important in the workings of the manifested world, from the causal standpoint they have a fixed order. The principle of change is the root cause of evolution of the manifested world. Without this principle there could not have been any movement in the Absolute and so the manifested world could not have come into being. The other three principles are also derived from the principle of change. Had there been no change in the Absolute the other three principles could not have come into existence. The world we live in is conceived by the Absolute through an exercise of its will. The Absolute could have existed even without manifesting a bulk of energy and through it the world. The Absolute willed to manifest energy and to evolve the world in a design pre-determined by it. The manifested world is only the fulfilment of a purpose of the Absolute. The principle of manifestation comes next. It is an outcome of the principle of will of the Absolute. The Absolute wills to manifest the world and actually manifests it. The evolved world moves on, following this principle of manifestation. The last principle in the above series is the group of scientific laws. These are laws derived from the principle of manifestation and from the principles inherent in energy. These laws are also the outcome of the principles of change and the will of the Absolute. The scientific laws are practically the joint effect of the first three principles of the above order. The scientific laws are all derived from the principles of change, will of the Absolute and manifestation.

Change is the foundational principle behind the evolution of the manifested world. It is the principle that works behind the principles of will, manifestation and scientific laws. The principles of will of the Absolute, manifestation and the scientific laws are only the specialized forms of the principle of change. The Absolute wills and wills effectively to bring about a change in the existing order. The will of the Absolute remains ineffective if the principle of change does not work behind it. The will of the Absolute, which itself is a change, brings about a further change in the changing order of things. The principle of manifestation also implies change in it. Manifestation is the transformation of an implicit entity to its explicit form. This transformation is not possible if the principle of change does not work behind it. Manifestation is thus a particular form of change. Lastly, the scientific laws are all principles that govern the processes of movement. If the manifested world stands still forever the said laws become inoperative. Even those laws of science which are related to stationary objects assume movements behind those objects and explain their stationary position as only an equilibrium of various forces working on them. The scientific laws are thus laws derived from the principle of change, or, in other words, it may be said that the scientific laws are only specialized forms of the principle of change.

The will of the Absolute is the direct cause of the principle of manifestation of energy. The principle of manifestation, though logically implied in the principle of change, does not necessarily follow it. It is the will of the Absolute that diverts the principle of change towards the principle of manifestation of energy. Similarly, the will of the Absolute is under no compulsion to manifest energy and evolve the world. Manifestation of energy is the actualization of one of the many courses that the Absolute can will. The will of the Absolute reaches the stage of its fruition in the scientific laws. The will executes its plan and fulfils its purpose through the scientific laws. The scientific laws had been formulated by the Absolute in such a way that their workings on manifested energy conform to the purposes, willed by it. The collocation of manifested energy at a particular stage may seem to be chaotic but that chaos is only a stage of movement towards a cosmos planned by the will of the Absolute.

The principle of manifestation is the foundation for the scientific laws. The said laws, according to us, are the detailed workings of the principle of manifestation. The principle of manifestation is the very postulate for all the scientific laws. The manifestation of energy necessitates the formulation of scientific laws. The behaviour of energy in their implicit form is not governed by the scientific laws. The behaviour of energy in its implicit form is the resultant of the workings of time or change, causality and consciousness on it. But the behaviour of implicit energy undergoes a radical change when it assumes an explicit form. Due to its manifestation a new element, the principle of space, now works on it resulting in the said change in its behaviour. The scientific laws are resultants of the composition of the principles of energy, time, causality, consciousness and space. Science deals only with the entities which are manifested and as such the scientific laws are the direct outcome of the principle of manifestation.

In every scientific law, change or time is involved. All manifested entities are in time and time is a postulate for all scientific laws. An object which is not in time is beyond the jurisdiction of science. An object of our consciousness which is conceived as "timeless" (by "timeless" we here mean successionless and not durationless) is not governed by the scientific laws. For example, the concepts of art, morality and religion are "timeless" and not governed by the scientific laws. Now, any law, not to speak of the scientific laws only, implies a purpose behind it. A law implies a cause behind it, a cause that works behind the formulation of the law. Behind every law there is at least the purpose of bringing certain phenomena under the jurisdiction of that law. Now, a purpose implies a will behind it, because the purpose is conceived in the will. Again, a law implies that it will be conformed to by certain phenomena. This implication of the law also means that there is a purpose behind every law. And a purpose, according to us, is conceived in a will. The scientific laws thus imply the exercise of a will behind them.

The principle of manifestation stands as the postulate for all scientific laws. Space is brought out by this principle and the scientific laws deal with entities which are spatial. The non-spatial and non-temporal concepts of pure mathematics and deductions from them are not scientific in the strict sense of the term. These are only deductions, logical. The laws in applied mathematics deal with entities that are spatio-temporal and as such these laws are scientific in the true sense of the term.

The scientific laws govern only the entities that are spatio-temporal and as such the principle of manifestation is involved in them. The principle of manifestation makes the entities assume a bounded form. Having a boundary of its own each entity will be separated from other such entities and will keep itself at a distance from other entities. The said boundary and distance are the very concepts of space. The scientific laws that imply the existence of space thus implies the principle of manifestation. To sum up, the laws of science are the resultants of the principles of time or change, will of the Absolute and space or the principle of manifestation.

Let us now see what the effects of different principles and laws on the bulk of energy, manifested by the Absolute, are. The principles of change and time make energy an ever-dynamic entity. Energy, or its configured form, matter, is undergoing change every moment. Nature, which is spatial energy transformed, is not stable anywhere. Plants and animals are born; they grow up and ultimately die. Land and rocks, seas and rivers, air and atmosphere, stars and planets are undergoing change every moment. Nothing in Nature remains what it was a moment earlier. The objects of Nature that seem to be changeless are also not actually so. Every object in Nature is composed of a cluster of atoms. The nucleus of each atom is composed of protons, neutrons, positrons, masotrons, masons etc. (or energy in the form of positive charges of electricity) and electrons (or energy in the form of negative charges of electricity) in various numbers revolve round this nucleus in different orbits. There is an equilibrium of electric charges in a stable atom. This equilibrium is the resultant of movement and countermovement of electric charges in the atom. Though the atoms maintain a state of equilibrium, there are movements and countermovements in them. Thus, the objects of Nature, or manifested energy, are dynamic at their core. Time or change is thus involved in the very existence of manifested energy.

The will of the Absolute makes manifested energy move towards the fulfilment of a purpose. In course of movements, energy assumes a comparatively stable form in the atoms of elements. Atoms of different elements combine together to form compounds and some compounds of the elements, carbon, hydrogen and oxygen in a favourable condition ultimately evolve life in them. One may here point out that in view of the fact that life evolves only in compounds of carbon, hydrogen and oxygen it is unwarranted to make the conclusion that the other purpose behind the movements of other elements also is to evolve life. Our explanation here will be that life not only means its emergence but it also means its sustenance. And for the sustenance of life a favourable environment is necessary. Elements, besides carbon, hydrogen and oxygen are necessary for the creation of that environment. Thus these elements contribute either directly or indirectly towards the evolution and sustenance of life. Some of the physical elements are necessary for the constitution of organism and some are necessary for the sustenance of those organisms. The changes in course of millions of years transform even the hard rocks into sand and soil and create conditions favourable for the evolution of life. Due to astral explosions and changes in temperature the atoms of physical elements undergo changes and new elements are formed. Transformation of the elements create favourable environment for some compounds of carbon to evolve life in them. Even the dead stars have the potentialities of evolving life because they are also constituted by the corpuscles of energy the collocation of which may undergo transformation creating a favourable environment for the emergence of life. Now, a striking fact that may be observed in the emergence of life is that life emerges in some carbon compounds only after the creation of a favourable environment for its sustenance. By "a favourable environment" we mean an optimum temperature in a suitable atmosphere and the provision for food, water, etc. which are necessary for the sustenance of organisms. It shows that the emergence of life and conditions for its continuance are parts of the same plan.

Water evolves before the emergence of organisms. Plants evolve before the emergence of herbivorous insects and animals. The herbivores emerge before the carnivores. All these facts point out that they fit in a scheme planned by the Absolute. If these facts would have been the results of mechanical forces only with no plan or purpose behind them, then there could be found instances in the opposite order. Animals would have come into existence before the emergence of their food and a suitable environmental condition. It may be said that the emergence of organisms is only a later stage in the mechanical transformation of energy than those of their food and the environmental condition. The evolution of air and water is earlier than those of organisms because the latter is a later product of the mechanical transformation of energy. Owing only to mechanical factors, when water and air attain a particular range of temperature, the carbon compounds evolve organisms. Even if we accept this explanation a question will still remain unanswered. Why was the creation of food earlier than the evolution of organisms who were to consume that food for their sustenance? Why did the organisms who could absorb carbon from Nature and could convert it to organic compounds evolve earlier than those organisms who could absorb carbon from Nature and could convert it to organic compounds evolve earlier than those organisms who could not do so but had to depend on the former for their sustenance? According to us, these phenomena of carbon fixation are clear indications of the fact that a will works behind the evolution of organisms. Had the evolution of plants (who can absorb carbon from Nature and turn it to organic compounds) and animals (who can only collect carbon compounds from the plant kingdom) been determined by mechanical reasons only, animals could have evolved earlier than those of the plants. If the evolution of organisms was due to mechanical reasons only, then any organism could have come into existence if only the mechanical conditions were fulfilled. How could the evolution of organisms be connected with the question of their food? Evolution of an organism due only to mechanical reasons does not necessarily exclude the possibility of its destruction for want of food. Why should food be provided first so that the organisms that emerge later can consume that food. If the courses of evolution are guided by the mechanical forces only, then the evolution of an organism and the emergence of its food cannot have a connection. But there is such a connection found in every case of the evolution of organisms. It may here be mentioned that overpopulation of a particular species of organisms and shortage of its food is a separate issue. Overpopulation and shortage of its food is determined not only by Nature but by the organisms themselves. As such this phenomenon is beyond the scope of our discussion. We are concerned with the evolution of the first species of organisms and not with their population due to reproduction in course of years. The biological developments that take place in the body of a mother before the birth of her child also show that Nature provides food before the oncoming of an organism. There are many more instances of this kind which show that a will works behind the movements of manifested energy. Had the evolution of organisms been determined by mechanical forces only, the necessity of food for those organisms could not have been envisaged beforehand.

The kingdom of inorganic objects is connected with the kingdom of organic objects through the phenomena of carbon fixation in organic bodies. The environmental conditions necessary for the growth and development of organisms are also provided by the inorganic objects of nature. It may therefore be assumed that the will that works behind the evolution of organisms also work behind the evolution of inorganic objects, though in a subtle manner, that cannot be apprehended easily. It is, therefore, not too anthropomorphic to say that the inorganic world acts as a background for the growth and development of the organic world. It may also be said that the movements and developments in the inorganic world is oriented towards the growth and development of organisms because the latter is a later stage in the scale of evolution of the world as a whole. The will of the Absolute thus works behind the change and movements of manifested energy.

The very term "manifested energy" means that the principles of manifestation work on such energy at every stage of its existence. The principle of manifestation, according to us, is the principle behind separation of entities from each other, a principle of differentiation that tends to keep each entity within a boundary of its own. Manifestation of energy at the very first stage makes it to separate itself from the integrated mass of other entities in the Absolute. It does not of course disconnect itself from the Absolute. It does not of course disconnect itself from the Absolute, because the latter has no boundary. It only assumes a form, well defined, within the Absolute. Manifested energy is now an entity, distinguishable but not separable from the Absolute. Only after assuming a specific form of its own and thereby differentiating itself from the Absolute, the bulk of manifested energy cannot cease to undergo change because the principle of manifestation is still working on it. The principle of manifestation makes this differentiated energy divide itself into corpuscles of energy so that each indivisible unit of such energy may differentiate itself from other such units and may assume a bounded form. These corpuscles of energy constitute the stuff behind this material world and the sciences deal with them. As the principle of time or change still operates on the said corpuscles of energy, a movement is generated in them and they therefore move about in space (which has already come into existence to accommodate the bulk of manifested energy). The laws of this movement are specific for the manifested form of energy. These are laws of energy by virtue of the fact that the said energy has assumed a manifested form.

On account of movements in the corpuscles of energy, they come in contact with each other and their compounds are formed. These compounds may be electrons, protons etc. or other units of energy even smaller than them. As the electrons, protons etc. are not identical in their constitution, it may be assumed that they are not the smallest and indivisible units of energy. The indivisible units are likely to be identical in nature, which the scientists may discover in future. Differentiation of energy into its corpuscles means that the corpuscles are separated one from another. Being separated from one another each such corpuscle must have something specific of its own, though all the corpuscles may have the same nature. To summarise, the processes of differentiation (due to manifestation) and integration (due to change or movement) affects the bulk of manifested energy. Due to differentiation manifested energy is converted into its corpuscles. Due to integration these corpuscles form compounds which may either be the electrons, protons etc. or compounds even smaller than them, which ultimately form the electrons, protons etc. Owing now to the principles of differentiation working on them, the electrons, protons etc. differentiate themselves from each other. Each one of such entities wants to maintain its integrity and independence. The specific structure or constitution of each such entity is the outcome of the workings of the laws specific for energy in movement. The laws pertaining to movement in general coupled with the laws inherent in manifested energy make the electrons, protons etc. assume their specific structure. Now the principle of change goes on working on the electrons, protons etc. and makes them to move about in accordance with the laws of energy in movement. The electrons, protons etc. therefore come in contact with each other and form compounds. Owing to workings of the principle of differentiation the electrons, protons etc. try to keep their integrity intact as also to absorb entities from outside. The said phenomenon of absorption is the principle of integration. Each one of the electrons, protons etc. tries to absorb other such entities in itself and not to be absorbed in others. The force to resist the possibility of being absorbed in others automatically makes others being absorbed. The principles of differentiation and integration therefore generate a tension in the kingdom of electrons, protons etc. and a continuous struggle ensues herein. But sometimes a balance is reached when some of them are forced to be absorbed in other such entities. Such a stable form is there in the structures known as the atoms. In an atom the electrons, protons etc. have come to a comparatively stable form due to the forces and counter-forces operating in between them. The atoms are of various kinds as the number of electrons, protons etc. in them and the consequent forces and counter-forces are different. Any number of electrons, protons etc. cannot, of course, come to a stable form and constitute atoms. Only some combinations of electrons, protons etc. reach at a balance of forces and counter-forces and become stable. But this stability does not mean that the electrons, protons etc. lie inactive. They cannot be inactive because activity or movement is their very essence. They go on exerting forces towards each other even when they maintain an equilibrium. This equilibrium is actually a tension or arrested movement of energy contained in the electrons, protons etc. As such, when this equilibrium is lost, some or all the electrons, protons etc. get loose and the arrested movement in them is released. Thus, due to workings of the principles of differentiation and integration on them, the electrons, protons etc. try to maintain their independence and integrity, fail to maintain them when they form atoms, and try incessantly to regain them again. The principle of integration makes the electrons, protons etc. to form atoms but the principles of change and differentiation continue to work on these atoms. Due to the principle of differentiation working on it, an atom tries to keep its integrity intact or to maintain its structure, already acquired. The principle of change makes an atom move about in space and come in contact with other atoms. The laws specific for energy also work on the movement of atoms and the whole process is oriented towards the fulfilment of will of the Absolute. The joint effect of all the said principles is the formation of molecules, or the clusters of atoms, and their compounds. As in the case of atoms, the molecules also move, come in contact with each other and ultimately make the pure elements like iron, silver, hydrogen, oxygen etc. and their compounds. The same forces that make elements and their compounds also break them up to form new elements and new compounds. But this last process cannot go on forever without producing anything, because, in that case the principle of will of the Absolute will be futile. The principle of will of the Absolute would, in that case, be ineffective because the phenomena of formation and breaking up of elements and their compounds serve no purpose of the Absolute. The principle of manifestation (or differentiation) is to make the entities, on which it operates, independent and free, for that is the exact meaning of differentiation. But if the said elements and their compounds only go on breaking themselves up to form new elements and compounds then they fail to be independent and free. They fail to have any self-contained, free and exclusive lives of their own. In other words, they fail to develop individuality which they should do if the principle of manifestation is fully operative on them. Due to full manifestation, a fully differentiated entity develops all its features fully. In the case of an unmanifested entity its features are not distinguishable because the principle of differentiation does not operate on it. Manifestation brings out the specific features of an entity and the entity starts living an exclusive life of its own. In other words, the entity becomes an individual.

Due to workings of the principle of change, the will of the Absolute, manifestation and the specific laws of energy, the elements and their compounds show tendencies to form entities of the nature of an individual. And the first step towards that end can only be the formation of a particular element or a particular compound of elements which may in due course develop into an individual or individuals. We find such an entity in some compounds of carbon, hydrogen and oxygen which, according to the bio-chemists, ultimately manifest life. Life, according to us, is a principle that has features exclusive to itself. It is a self-contained principle, though not completely free. It evolves out of matter but neither takes matter into it nor gives matter out of it. It is thus a self-contained principle having exclusiveness in its nature. It cannot be equated with matter nor can matter be equated with it. It cannot be made to form a compound with anything else, including even entities of its own kind. It has thus a peculiar exclusiveness in its nature. Can life be called an accidental phenomenon or can it be a by-product of matter? Our answers to both these questions are in the negative. No event is accidental in this universe because every event is determined by its antecedent conditions. Life evolves out of a particular compound of matter at a particular range of temperature and pressure. And it surely means that a particular compound of matter and a particular range of temperature and pressure are necessary for the evolution of life. The said compound of matter, temperature and pressure are the antecedent conditions for the appearance of life. How, therefore, can life be an accidental phenomenon? In the same sense life cannot also be a by-product of matter. It is rather the product of a particular compound of matter in a particular environmental condition. It if is asked, why only a particular compound of matter manifests life, our answer will be: because the Absolute so wills. We do not know why any matter or any compound of matter does not manifest life.

Life, as we have said earlier, is a principle that makes a compound of elements to keep it separated from other elements or their compounds. A compound of such nature tries to live an exclusive life of its own, and for this, it draws towards itself suitable materials from outside so as to keep this principle intact. It also thereby grows. The materials which are consumed by it as food change their characteristics radically and convert themselves to organic compounds. In other words, the food consumed by it also acquires the characteristics of life in it. Thus the transformation of inorganic elements or compounds into organic compounds is due to manifestation of the principle of life in the former. Not all the behaviour of an organic compound is governed by the so-called mechanical laws of science that govern the behaviour of inorganic elements and their compounds. Some of the said behaviours of an organic compound are governed by the laws of life also. The basic difference between these two sets of laws is that while the mechanical laws determine an event conclusively, the laws of life do not do it. Accurate prediction of an event is quite possible when only the mechanical laws are to be accounted for. There cannot be any element of probability in the mechanical laws because their fields of application, i.e. the inorganic elements and their compounds have no exclusive lives of their own. The inorganic elements and their compounds have no principle of exclusiveness working on them that can determine their behaviour. Absence of an inner and exclusive life in an inorganic compound makes all its behaviour determinable from outside. Probability, in the true sense of the term, or the presence of a number of possible courses of action, is not a characteristic of the mechanical laws. On the other hand, the laws that govern life are basically different in this respect. Life defies the mechanical laws because it is a self-contained principle that has a peculiar exclusiveness in its nature. Life can control itself from within and as such its behaviour cannot be determined from outside accurately. Though applicable to a certain extent the mechanical laws cannot determine the phenomena of life conclusively. In the phenomena of life, therefore, there is a possibility of alternative courses of behaviour which cannot be determined from outside accurately. Life has its own principles and it works from within. All its behaviour cannot therefore be determined from outside by applying the mechanical laws. The mechanical laws or the laws of science are laws of elements, or clusters of energy corpuscles that have not yet developed any principle of individuation. The said corpuscles of energy have attained a state of equilibrium by balancing their movements. They have formed a community through cohesion. But the forces they exert or can exert and their movements can be studied from outside, because, they have no other inner force, exclusively controlled by them, that cannot be determined from outside. But the organic compounds having life have developed a principle of individuation. This principle equips them with a capacity to control their behaviour from within. An inorganic compound endowed with this capacity is an organic compound. The transformation of compounds from inorganic to organic is, therefore, a further step towards manifestation. As such, life, which is a principle that controls itself from within and has an existence exlusive to itself, is a higher stage of manifestation than that of the manifestation of inorganic elements and their compounds.

Life is manifested only in some of the compounds of carbon. There is no reason other than the will of the Absolute that can be said to have determined the manifestation of life in a particular compound of carbon. We from outside can only discover laws that determine the behaviour of inorganic elements and their compounds. But in the case of organic compounds even these laws are not applicable because of the activities of life, or the exclusive inner principle. We, therefore, cannot find any reason why a particular compound of carbon manifests life and others not. This reason will be known only to the being who controls the inner and exclusive principle of individuation.

In other words, it is is known only to the absolute who knows and determines the manifestation of life. Thus, it is the will of the absolute that determines the manifestation of life in some components of carbon and also works behind it.

The principles of change, the will of the Absolute, the principle of manifestation and the scientific laws are in operation upto the evolution of inorganic elements and their compounds. But if there is no further evolution thereafter, if only the processes of formation, dissolution and re-formation of inorganic elements and their compounds go on forever, then the laws of change and the scientific laws will of course be in operation but the principles of will of the Absolute and manifestation will be inoperative. We have shown earlier that the principles of change and the scientific laws cannot remain inoperative as they are implied in the very manifestation of a bulk of energy from the Absolute. But in the case of the will of the Absolute and manifestation it is otherwise. The will of the Absolute works behind the manifestation of energy and formation of energy corpuscles, electrons-protons etc., atoms, molecules, elements and their compounds. The said entities come into being for the fulfilment of a purpose willed by the Absolute. But if at the stage of evolution of elements and their compounds there is only formation and re-formation of entities of the same kind, then they serve no purpose of the Absolute. They will serve no purpose because any one set of them will not be basically different from its preceding or succeeding set. As no entity of a different kind is coming into existence evolution of entities will come to a halt and as such there will not be any further manifestation of energy. Can it be said that formation, dissolution and re-formation of elements and their compounds were the only events willed by the Absolute? No, because it would then frustrate the very purpose of manifesting a bulk of energy from the Absolute. If, to break the monotonies involved in the lonely existence of the Absolute is the purpose for manifesting a bulk of energy, then, to exist lonely or to exist alongside a bulk of manifested energy in the process of formation, dissolution and re-formation of entities of the same kind frustrates that purpose. And the very act of the Absolute in manifesting a bulk of its energy will in that case be futile. We cannot conceive of the Absolute to have undertaken such a fruitless venture, because, then, the very conception of the Absolute will get frustrated. Moreover, the principle of manifestation, which means the principle of separation of entities from each other and their having lives exclusive to their own, will in that case be barren. The full play of the principle of manifestation on an entity means that the entity lives an exclusive life of its own and also maintains that form of its existence. But, if the elements and their compounds only undergo change following the scientific laws they cannot be said to have acquired the capacity of maintaining their exclusiveness and integrity because they do not remain identical throughout the process of that change. The principle of manifestation thus remains ineffective. The coming into existence of inorganic elements and their compounds cannot therefore be the last stage of evolution.

Therefore, at this stage of evolution, the principles of will of the Absolute and manifestation transform some inorganic compounds into an organic form and manifest life in them. Manifestation of life in some inorganic compounds is a step farther than the formation of elements and their compounds. In manifesting life the will of the Absolute is active again. Through organic compounds and life, the inorganic compounds move in a new direction and as such the will of the Absolute is on the move again and its continuity is maintained. Evolution of life also maintains continuity in the operation of the principle of manifestation. The principle of separation of entities from each other which started at the manifestation of a bulk of energy from the Absolute and worked through the formation of energy corpuscles, electrons-protons etc. continues to work through the formation of organic compounds and manifestation of life therein. The inorganic elements and their compounds had a separate existence of their own but they were not able to maintain that state of their existence because of the principle of change working on them vigorously and breaking them up to form new elements and new compounds. The evolution of organic compounds is a farther step of workings of the principle of manifestation because the organic compounds are at an advanced and developed stage of the principle of separation and exclusiveness in the sense that such compounds are farther separated from each other and have a more free, more exclusive and more individual existence of their own. But the organic compounds are also not able to maintain their integrity due to the principle of change working on them. At the evolution of life in the organic compounds the principle of individuation, which is implied in the principle of manifestation, reaches that stage where the evolved entity is able to maintain its integrity, independence and exclusiveness in spite of the principle of change working on it, because, to us, life is a principle that makes the living being concerned, live an exclusive life of its own. Thus, at the evolution of life, the principles of manifestation and the will of the Absolute are operative again. It may also thus be said that at the evolution of life, the principles of change, will of the Absolute, manifestation and the scientific laws continue equally to play their respective roles on the bulk of energy, released and manifested by the Absolute.

The organic compounds have inorganic constituents with only an extra principle, the principle of individuation added to them. The principle of individuation is an advanced and a later stage of the principle of manifestation. It assumes various forms and ultimately reaches the stage of organic compounds with life therein. Now, manifestation, which means presentation or exhibition, is itself a principle of consciousness. Consciousness is 'an act of being conscious of' or 'to be aware of' something presented before it. 'I am conscious of your presence' means that I am aware of the fact that you have presented yourself before me. Consciousness is the principle behind the act of being conscious of something presented before it. 'Presentation' is meaningless when no one is aware of it, when no one responds to it. The act of 'presentation' or 'exhibition' or manifestation thus implies the 'act of being conscious of', or the principle of consciousness. To present oneself, one is at least required to know that he is presenting something, i.e. every presentation or exhibition or manifestation implies an act of being conscious of it. Manifestation thus implies consciousness behind it.

In view of the above, when a bulk of energy is manifested by the Absolute that energy includes consciousness in its very existence. The corpuscles of energy, electrons-protons etc. are all conscious in this sense. But the consciousness of all the said entities is not of the same level. Along with the evolution of these entities one after the other, consciousness also evolves or undergoes development step by step. The corpuscles of energy have a more developed form of consciousness than that of a bulk of energy manifested by the Absolute, the electrons-protons etc. have a more developed form of consciousness than those of the corpuscles of energy and so on. When a bulk of energy reaches through evolution the stage of organic compounds with life therein, the principle of manifestation has almost assumed the form of individuation. According to us, the principle of manifestation in its fullest development is the principle of individuation. The earlier stages of the principle of manifestation are also conscious principles but only in a less developed and vague form. When a manifested entity is fully individuated, the consciousness in it will be in its fully developed form. An undeveloped form of consciousness is that consciousness which is so vague that it cannot even be differentiated from energy where it is in. And the most developed form of consciousness in an organic compound is a sharply differentiated principle.

Now, life, as we have shown above, is the principle of consciousness clearly manifested in organic compounds. With the manifestation of life in an organic compound, the said compound is sharply differentiated from other organic and inorganic compounds and the principle of individuation begins to operate. To make life to continue to exist, a suitable condition is to be maintained in the organic compounds. Due to workings of the principle of change on it, an organic compound goes on decomposing itself every moment. Therefore, in order to maintain the condition in the said compound more or less identical and suitable for life, materials from outside are to be taken in. Thus there arises the necessity of food, water, air etc. and also a particular range of temperature and pressure. If, however, a suitable condition is not maintained in the organic compounds, life evolved in it will have to depart from it.

Consciousness and life do implicitly exist in manifested energy and its evolutes up to the stage of inorganic compounds, but being not differentiated principles they do not exert influences in the movement of such evolutes. As such, the scientific laws are almost exhaustive in explaining the movement of electrons-protons etc., molecules, elements and the inorganic compounds. But with the evolution of consciousness and life in the organic compounds, the scientific laws do not remain so exhaustive. It is so because consciousness and life now control to a large extent the movements and behaviour of organic compounds from within and this controlling force, being an exclusive principle of individuation, cannot be studied from outside. Life thus defies all scientific laws because these laws do not cover the influences due to life, which is an individuated principle of exclusion.

After the evolution of life, the four principles—the principles of change, will of the Absolute, manifestation and the scientific laws, add a fifth principle to their lot to guide the course of evolution of life. This fifth principle is the principle of individuation or of life. Life has its own laws which work from within and cannot therefore be formulated like the laws of science. Being a principle manifested in energy, life is governed to a large extent by the laws of energy but the additional force of inner control exerted by life makes the organism and entity that cannot be studied in the scientific way from outside.

The process of manifestation does not terminate at the evolution of life because life has the possibility of developing further. Moreover, the five principles—the principles of change, will of the Absolute, manifestation, scientific laws and the laws of life, also go on acting on a living organism and make it develop further. Change, the very essence and law of existence, is there in the organic compound and in life manifested therein. An organic compound, like every other thing, undergoes change every moment and life also is a principle which implies change in it. Moreover, change, as the essence of existence, has no initial or final stage of it. The will of the Absolute worked behind the manifestation of energy and oriented its course towards the fulfilment of a purpose. At the manifestation of life a principle of individuation is formed but it is still in a vague form. All the movements of life at this early stage are blind and as such, the said movements serve no purpose of the Absolute. A clear principle of inner control needs therefore to be formed in life so that the process of life, instead of being only a blind movement may take a well-deliberated, clear and sharply defined course. If life continues to be a blind and haphazard movement, then it leads to nothing and serves no purpose of the Absolute. We shall have to say then that the Absolute had no end in view while manifesting a bulk of its energy. But to entertain the idea of Absolute involved in a fruitless venture is to destroy the very conception of an Absolute which has infinite consciousness in it. To make life a planned venture the Absolute must therefore conceive of a centre of inner control in it. This centre of inner control in life is the principle of a mind in it. Mind is a principle for deliberation and control over the movements of life.

Again manifestation of life implies its continuance and not its discontinuance or destruction. Now, life, as we have shown earlier, is a principle the existence of which depends on certain conditions in the organic compound in which it is manifested. To keep the torch of life in it burning an organism is in need of food to make good the loss in it due to change. By food we mean the organic and inorganic compounds that are to be consumed by the organism for the maintenance of its body. Water and air are also necessary for maintenance of its body. Another necessity for the continuance of life in an organism is a particular range of temperature and pressure. If these materials are not readily available the organism will then have to collect these materials. Even if these items are easily available the organism will at least have to consume them, according to its necessity. If the organism is mobile it may collect its food from different places and if it is immobile it will at least have to consume the food available in its immediate neighbourhood. In both the cases, some or all the activities of the organism are motivated towards the collection of its food and its movements towards a suitable environment. Some sense of apprehension, however feeble it may be, is thus involved in the activities of an organism. It may be said that the activities in question are only instinctive. Even when these activities are so a principle of apprehension works behind them so that the organism may maintain its body and may keep the flame of life burning in it. Being not a fruitless creation of the Absolute, life, once evolved by it, will be preserved. And for preservation of life in an organism, a sense of apprehension will have to be formed in it. A feeble form of this principle of apprehension is an instinct. When different instincts counteract or strengthen each other, a resultant automatically evolves and the organism moves accordingly. A central organ to control the movements of an organism is thus to be formed if the flame of life is to be kept burning. There must at least be two basic instincts working for the maintenance of life. One of them works behind the activities for collection of food etc. and the other works behind the activities in connection with protection of life from malicious forces. If in an organism these two instincts are made to work, simultaneously by circumstantial forces (for example, when the organism is placed in such a situation that it must collect its food but its activities on this account mean its exposure to malicious forces), then the organism will be in the horns of a dilemma. If it moves to collect food for maintenance of its body it will have to face death and if instead it tries to avoid being exposed to malicious forces it will have to die of starvation. Placed in such a situation, the organism will have to choose one of the two alternatives. We cannot say that the course of action the organism will adopt will be the numerical resultant of the two urges working in him, i.e. if the urge for food is stronger than the urge for protection of life then the organism will try to collect food and if it is otherwise, the organism will then postpone its collection of food. Here the term 'numerical resultant' may either mean that these two urges are mechanical forces that work against each other or that though these urges are not mechanical forces they at least mean a cancellation of one by the other. In both these cases, after cancelling the weaker urge, the stronger urge must be poorer in strength than what it was before. If the organism moves for collection of food, it will now move slowly. If on the other hand the organism tries to protect its life, it will do so less vigorously. But actually, an organism in such a situation acts in the opposite way. If the organism moves to collect its food, it does so hurriedly and if it tries to protect itself from the malicious forces, it does so more vigorously. Thus even the instinctive behaviour of an organism cannot be explained in terms of mechanical forces and so cannot be determined mathematically. A principle of deliberation is thus present in the activities of life in an organism. This principle of deliberation must have a centre for it where the different instinctive urges meet each other and are balanced. This centre in the life of an organism is the principle of mind. Thus, the evolution of life in an organism automatically implies the creation of a centre in it for deliberations etc. for its continuous existence and further development.

When the principle of manifestation works on life, the latter will have a tendency to move towards individuation, for that is the very meaning of manifestation. Now, the manifestation of life in an organic compound means that the principle of life is confined in that compound only. Being spatial in its very constitution, an organic compound can be individuated by a boundary. But life, being a principle of consciousness, cannot be individuated in that way. For individuation of the principle of life, therefore, a sense of 'mine' and 'thine' is to be generated. And the evolution of a sense of 'mine' and 'thine' means the evolution of a power of discrimination. But life is not only required to exercise its power of discrimination, it has its other functions too. Even when some discriminative activities go on in life, its other activities cannot be suspended. The discriminative activities must therefore be confined in a particular sphere of life leaving its other spheres engaged in other activities.

Mind, as the principle of discrimination, will first discriminate itself from the organic compound to which it is attached. In other words, it will be conscious of its own extent. Mind will then discriminate itself from every other entity, including other minds. It will also be conscious of the life of which it is the central organ and will develop a sense of owning it. The mind becomes aware of the fact that the changes in the body to which it is attached are transmitted to it and also that its own changes are transmitted to the body. The mind thus comes to know that the said body is its sphere of influence and that its own fate is correlated to the fate of the body. The sense of owning the body makes the mind identify itself with the body. The mind-body complex being complete, the mind considers all the requirements of the body as its own and counsels for or even guides the behaviour of the body.

The evolution of organic compounds, life and mind do imply that they are to continue to exist because their existence contains no counter-principle in them to make them to discontinue to exist. Now, if the organic compound is to continue to exist, it will have to retain the characteristics evolved in it. To retain its characteristics, the organic compound will have to behave in a particular manner and such behaviour will have their effects on life and mind in the forms of reflections on them. These are behaviour or functions specific for the organic compound and as such cannot be guided or controlled by the mind. The mind can only know them retrospectively. But as the mind has a sense of owning the body, it will try to guide the body whenever there arises conflicts between two or more instincts. Like the body, the life and mind also have their specific functions for continued existence and progress. The principle of discrimination in mind makes it come to know that its continued existence and progress depend on the welfare of the body. The mind therefore tries to guide all activities of the body, so as to meet its own requirements. In every critical situation when the body is threatened with possibilities of death and destruction, the mind comes forward to interrogate the situation, to find out means for the avoidance of danger and to guide the body in its activities. Due to its urge for continued existence, life, through its central organ, the mind, looks after the welfare of the body. Being a principle of discrimination the mind has also its power of apprehension. It goes on analyzing and discriminating things and apprehends the situations. For this specific function of life and mind also, a particular bodily condition is necessary. The mind directs and guides the body to maintain that condition also. As principles of consciousness life and mind must have their own ways of coming in contact with other entities. But as at the initial stage of their evolution their power of discrimination is not developed, they cannot easily differentiate themselves from the body in which they are evolved. At a later and developed stage, the mind comes to know that it itself and life are principles of consciousness and not of matter or energy. The mind comes also to know that it has its own ways of coming in contact with things and apprehending them. However, at the initial stage of its manifestation, the mind works through the body and comes in contact with the world through the body. So, for its own specific functions of discrimination and apprehension, the mind requires the body to be kept in a condition suitable for its own purposes. The mind therefore guides the body to make it maintain that condition.

Now, the specific function of mind is to discriminate. After discriminating or differentiating itself from life (of which it is the central organ) and body the mind integrates this knowledge by apprehension. To differentiate one entity from another is somehow to know something about the said entities and to place that knowledge in one's sphere of apprehension. While differentiating an entity from itself, therefore, the mind integrates its knowledge about that entity by placing that knowledge in its sphere of apprehension. As at the initial stages of its progress the mind works through the body it can have only an external or superficial view of things it comes in touch with and as such cannot know the real nature and various aspects of that entity. The mind's apprehension of the entity in question may therefore be incomplete, imperfect or even wrong. But on the basis of this apprehension the mind chalks out its plan of action and guides the body to act accordingly. On some occasion therefore the results of the action are not what were expected. The idea of the entity, apprehended by the mind, therefore needs some modification. The mind corrects or modifies its ideas of the entity and places the modified idea in its sphere of apprehension. The sphere of apprehension of the mind is thus enlarged every moment due to integration of ideas, modified ideas and corrected ideas of entities. Mind, which is a principle of consciousness, absorbs in this manner new materials every moment or rather enlarges itself by absorbing or integrating new entities every moment and makes progress towards the attainment of a state of infinite consciousness.

The evolution of life not only implies the evolution of a mind but also a will in it. The instinctive activities of the body are not within the control of the mind. The mind only comes to know them retrospectively. Being a form of consciousness, mind has also to abide by the general principles of consciousness. Mind, as a form of the principle of consciousness, has a specific course of activities which mind, as an individuated form of consciousness, can neither defy nor control. But as an individuated entity, the mind has its own sphere of influence and control. In all cases where deliberation and discrimination are necessary to determine the course of activities of the body and of the mind, the mind, as an individuated entity, chooses one out of two or many courses and thus exercises a will. This will is the will of an individuated form of consciousness and as such it is the individual will. Not being the will of an omniscient being, like the will of the Absolute, the individual will may commit errors and it frequently does so, particularly at the earlier stages of its evolution.

The work of the principle of manifestation is not complete at the stage when the mind only goes on coming in touch with entities other than in its body-life-mind complex and integrates its ideas of them in its sphere of apprehension, thus only enlarging its principle of consciousness. In course of gathering experiences, the mind goes on learning new lessons for its future activities. The mind learns that sometimes its calculations fail to bring expected results. The mind therefore tries to rectify the defects in its calculations and modifies its apprehension and knowledge of the entities in question accordingly. Thus ensues a fresh series of discrimination, deliberation and calculation in the mind. But the results of all these calculations and recalculations of the mind may mean only a slight improvement in its deliberations and a little better life for the organism. The miscalculations and wrong deliberations of the mind, as mentioned above, are all due to the fact that the mind works and gathers knowledge through the body. Unless the mind learns to work independently it cannot gather faultless knowledge of the world and attain the final stage of its development. If, to gather knowledge is the function of the mind, then to acquire that state when it can gather faultless and absolute knowledge will be the final stage of development for the mind. To work independently and to gather absolute knowledge the mind must discover its own inherent laws and act accordingly. And this state of the mind will be the highest stage of its development and the final stage of its individuation.

The body of an organism is a specific organization of energy. The inherent laws of energy therefore guide the behaviour of the body. But the life and mind of the organism are principles of consciousness and as such the inherent laws of life and mind are basically different from the laws of the body. Full manifestation and individuation of the mind is not therefore possible unless the mind discovers the specific laws of consciousness and acts accordingly. It may here be asked whether or not the mind moves automatically to follow the above course as the principle of manifestation is still working on it. Our answer to this question is as follows. The principle of manifestation works on energy and as a result there evolves the corpuscles of energy, electrons, protons etc., atoms, molecules, elements and their compounds. Due also to the principle of manifestation working on the compounds of elements, life and mind evolve in some such compounds. Up to this stage of evolution the principle of manifestation reigns supreme and the evolutes come into existence automatically. But with the evolution of mind and the principle of discrimination an individual will comes into existence to exercise its choice over alternatives of action. As a result the principle of manifestation does no longer reign supreme over the activities of an organism. Spontaneity in the work of the principle of manifestation being lost, the mind and body of an organism can no more make progress spontaneously. There cannot now be an automatic growth in its body and mind. The individual will being active, the growth of the mind is now voluntary. As such, the mind of an organism cannot discover automatically the inherent laws of consciousness. A mind can discover the inherent laws of consciousness only at its own initiative.

Now, the life and mind of an organism are manifested forms of that consciousness, which was implied and contained in the energy constituting its body. Though consciousness, as a principle, cannot be confined in a sheath made of energy, yet its connection with particular quanta of energy makes it limited. The consciousness of an organism is a limited entity also in the sense that it is not the consciousness of another organism. Consciousness, in the form of life and mind, evolves in an organic body and as such the first entity they are in touch with is that organic body. The mind starts functioning with the discrimination of entities, deliberates over their relations with each other and their possible effects on the organic body to which it is attached. It selects a line of action for the said body and accordingly guides it in its activities. In fact, a mind starts learning the art of discrimination through its said activities. Thus as its working ground the body is necessary for the development of the mind. It is, therefore, not unnatural that a mind keeps itself engaged in works for the body only and seldom pays attention to itself. It thinks for the welfare of the body, it feels the ups and downs of the body and it wills for the future activities of the body. It considers the limitations of the body as its own. It laments when the body suffers and it rejoices when the body is in a perfect condition.

The contemplative activities of the mind begin with its work of discrimination for the body. It then contemplates over entities other than the body which have or may have their effects on the body. In its third phase of contemplation, the mind keeps itself busy with matters more general, matters that affect the body, but only indirectly. These are affairs of the family, the community, the society and the state. In its fourth phase the mind is engaged in matters which seldom affect the body. These are contemplations relating to pure mathematics and the sciences, the principles of logic, the literary arts and the fine arts like painting, sculpture, music etc. Here the mind leaves the play of energy behind and enters into its own sphere of contemplation. In its fifth and the final phase, the mind enters into the domain of philosophy and religion and ultimately reaches its destination. The mind here begins to enquire about itself and the body to which it is attached and tries to apprehend the relation of this body-life-mind complex to the rest of the universe. Through its enquiries the mind comes to know that its inner being is neither the body to which it is attached nor any other physical object of the universe. The mind also comes to know that as a principle of consciousness it has the potentiality of being conscious of the whole universe and be one with the infinite consciousness of the Absolute. In its contemplation the mind not only discriminates entities from each other but also integrates the knowledge of their real nature and in so doing enlarges its sphere of apprehension. In other words, the mind, in so doing, proceeds step by step towards its attainment of infinite consciousness. But this is only a general and theoretical principle that works behind the mind's journey within. While actually undertaking the task of exploring itself, the mind faces hurdles of various types. In course of apprehending entities of the universe and analyzing their characters, the mind learns that it can, from outside, know only the phenomenal character of a thing. It can know the real nature of a thing, or the thing-in-itself only if it can be one with that thing. The mind comes in touch with a thing through its own framework and as such it can have only a distorted view of reality. In order to apprehend perfectly the objects of the universe the mind is required to know its own nature first so that it may develop a technique of defying its own framework and thus come in touch with objects directly.

The mind, as we have said earlier, is naturally attracted towards the physical objects because the latter are in easy touch with the former through the sense organs. It is, therefore, a hard task for the mind to revert its natural movement and to direct it towards itself. The mind has to undergo a hard discipline so as to be able to focus all its attention to itself. In course of this the mind learns the technique of making itself one with the objects of the universe so that it may apprehend their real nature. In this way the mind ultimately attains infinite consciousness and becomes one with the Absolute. As life and mind evolve in an organic body, the said body acts as their support till the mind becomes one with the Absolute. When at death the concerned life and mind leaves the body, the body loses its specific organization and becomes a compound of elements like other compounds. Though consciousness that evolved in the body was implicitly present in the latter it did not exhaust all consciousness that could be implied by the body because consciousness is an infinite and indivisible entity. Had it not been so, then the life and mind of an organism would have been large or small according as the organism was large or small. The life and mind of an elephant could, in that case, be larger than the life and mind of a man. This is something absurd in view of the fact that consciousness has no size like that of a body. Consciousness may have only limitations in different degrees due to its connections with organic material bodies. The mind of an organism is entangled in the affairs of the body and as such the limitations of the body limit the activities of the mind. The extent of mind's entanglement in the affairs of the body determines the degree of mind's limitations. Unless the said limitations are crossed over by the mind, it cannot attain that state when it can acquire absolute knowledge of the universe and can be one with the Absolute.

Mind has different phases of its progress towards attainment of infinitude but the said progress is made by the mind through its own efforts. As a principle of individuation, the mind has its own will which directs and guides its movements. The progress of the mind in its journey towards the Absolute depends on its individual will as also on the will of the Absolute which is active in the background. If its cravings to be one with the Absolute is too deep then a mind may overstep one or more of the said phases and may reach the final stage of its journey to attain infinite consciousness. If the mind can suspend activities of its individual will, then the will of the Absolute comes forward and functions in its place. Individual will, the organ for the principle of individuation directs the activities of the mind. A suspension of this will cannot stop the activities of the mind which is ever-active. Due to suspension of the individual will therefore a similar principle of will must act in its place. Being an entity individuated and exclusive to itself, the mind cannot accommodate the will of any entity other than that of the Absolute. In a case of suspension of the individual will, therefore, the will of the Absolute that was working from the background will come forward, take the place of individual will in the mind and will direct the activities of the mind. If the mind acts under the guidance of the Absolute, its apprehension will be perfect and its progress towards the attainment of infinitude will be rapid and sure. When at last the mind's apprehension of the universe is complete it becomes one with the infinite consciousness of the Absolute.

[A note on temperature and pressure:- By the words "temperature" and "pressure" we mean just what the physicists mean. "Temperature", to us, is a form of energy radiated due to inner workings of the elements. By "pressure" we mean the force exerted by elements or their compounds.]

Chapter 5

Principles Behind Evolution of the World

By evolution we mean the series of changes that take place in a logically connected sequence. It is a slow process of gradual change and development. The universe, as we find it now, is the result of evolutionary changes on a primal stuff. The sun, the moon, the stars, our earth with plants and animals on it did not come into being all at once. These entities have come to their present form in a gradual process of change and development which started on a primal stuff. We apply the principle of evolution to the manifested form of that primal stuff only because the implicit form of that stuff was obviously something to which the said principle cannot be applied.

Now, the source of this universe, as we have shown earlier, is the Absolute, which is an entity that connotes infinite consciousness coupled with infinite energy in it. The infinite consciousness of the Absolute is ever-manifested, because it cannot be otherwise, and it reflects the whole universe, leaving nothing beyond it. But the Absolute may not manifest its infinite energy all at once. Not being limited by anything the Absolute is omnipotent and omniscient. The Absolute cannot have any feeling of want for anything, for he contains all. He may continue to exist for ever without bringing our material world into existence. But due to reasons best known to him, the Absolute has brought this material world into existence. We do not find any logical necessity behind this act of the Absolute. Being an entity complete in itself, the Absolute cannot have any feeling of want for anything. All entities that are or may be are in the Absolute. Even if the material world is not brought into existence the Absolute may continue to exist but in that case he is bound to experience a monotonous existence. There is no logical necessity for the Absolute to avoid this monotony. But if the Absolute intends to overcome this monotony, what can he do? By rearranging his infinite energy and infinite consciousness in any manner, the Absolute cannot overcome the monotony because such arrangement cannot bring out anything unforeknown to him. Everything explicit or implicit, every form or arrangement, potential or actual, are known to the Absolute. To overcome the monotony, therefore, the Absolute will have to bring some entities into existence which may have the freedom to determine their own courses of action and such courses of action will not obviously be foreknown to the Absolute. The Absolute will thus be able to get some novel situations so that he may overcome the monotony in his existence. Though such an entity cannot have full independence in the exact sense of the term, it can certainly have a relative freedom in some spheres of its existence. Though the very existence of such an entity depends on the Absolute, it may have a sphere of its own.

According to us, an entity of that kind will be an image of the Absolute in the sense that it will enjoy, like the Absolute, a freedom in the exercise of its will to determine its state of existence. The Absolute will therefore bring a principle of will into existence which may function like his own will. The creation of a will necessitates the creation of a life and a mind. Now, life is a form of consciousness and it is also a principle of individuation. Consciousness, as we have shown earlier, implies energy in it, the latter being the principle of sustenance for the former. To evolve life, mind and will, therefore, the Absolute is to manifest energy and to orient its movements towards the evolution of life, which means the evolution of a principle of individuation, a sense of exclusiveness and privacy. All that is necessary for the sustenance of life need also to be evolved for the purpose. To overcome the monotony of his existence the Absolute is therefore to start working on a bulk of energy and to direct its movements towards the evolution of entities necessary for the sustenance of life, which is to appear at a later stage. When all these preparations are completed, life appears and undergoes development towards the evolution of mind and will. The above analysis may seem to be an explanation too anthropomorphic in character. The explanation seems to display a spirit of self-importance when we say that the sun, the moon and the stars, the mountains, the rivers and the seas, air, water and warmth are all meant for us, the living beings. We only beg here to mention that an explanation offered by a human being cannot but have a human element in it.

To the question why the Absolute brought our material world into existence the answer of some thinkers is that the creation of this world is only a sport for him. Let us examine this explanation and see whether or not it is acceptable to us. As there is no logical necessity behind the evolution of our material world, it is, in a figurative sense, the effect of the sportive activities of the Absolute. But in the strict sense of the term it cannot be a matter of sports for the Absolute, because in that case the Absolute will not be able to overcome the monotony of his existence. May it be a matter of sport or anything else, the products of evolution cannot be unknown to the Absolute. Moreover, as products of the sportive activities of the Absolute, the mind and will of living beings cannot have any freedom to determine the courses of their activities. The Absolute, in that case, will be responsible for the sorrows and sufferings of such beings. To break the monotony of his existence, the Absolute must therefore evolve such a principle that may work independently. Though not absolutely independent, such a principle will display something unforeknown to the Absolute. This novelty will break the monotony of his existence as also it will be a real sport for him. We thus go beyond the explanation examined here and say that for the sport to be made sport-like, the Absolute will have to evolve a principle of freedom of the will in living beings. The very purpose of evolving this material world is therefore to create individuated entities who, in due course, are to develop a freedom in their will.

To bring individuated entities into being, the Absolute starts, according to us, with the principle of manifestation. He manifests a bulk of his energy which was in him in an unmanifested form. Being now separated from the integrated mass of infinite energy of the Absolute, the said bulk of energy assumes a new form. This separation from the integrated mass does not of course mean that the bulk of energy has now a fully independent existence of its own. Manifestation of a bulk of energy does not mean a complete severance of its ties with the integrated mass of infinite energy of the Absolute. It does only mean the difference of its form from the form of its original state in the Absolute. It does also mean the assumption of a specific form of its own which makes it distinguishable from the integrated mass of energy in the Absolute. When the principle of manifestation works on the bulk of energy, the latter assumes at the first stage an explicit form and becomes a single mass of energy, its various elements being indistinguishable from each other. The principle of manifestation now works on this mass of explicit energy and makes its innumerable constituents to form corpuscles of energy, for that is the meaning of manifestation. By the word 'corpuscle' we mean the smallest unit of energy which is indivisible and which contains a single constituent in it. Each corpuscle of energy assumes a specific form of its own and is separated from other corpuscles so that each corpuscle may be individuated. The corpuscles of same or similar nature of will now try to form communities of themselves so that they may have a form separable and distinguishable from other communities of dissimilar nature. Thus, due to manifestation, each corpuscle and each group of similar corpuscles are individuated and through this process of differentiation and integration electrons, protons and other such entities are formed. Due to the fact that change and movement are the very essences of existence the electrons, protons etc. will be in movement and will come in contact with each other. The forces generated in the corpuscles of energy due to their tendencies of differentiating themselves from each other are balanced when they form communities in the form of electrons, protons etc. or even units smaller than them. When the electrons, protons or such like units of energy come in contact with each other, there ensues a play of forces and counter-forces due to tendencies of such units to keep their integrity intact. The forces of such units overlap each other's area of influence and ultimately come to a comparatively stable form, which are the atoms. The said play of forces and counter-forces may also end in diffusion of energy in the form of heat etc. In the same process of differentiation and integration, the atoms of same or similar nature come together to form a community and differentiate their kind from other kinds of such communities. When, due to movements, the elements of different kinds come in contact with each other the forces of differentiation and integration act and react on each other. This action and reaction will be hastened if diffused energy in the form of heat, pressure etc. also takes part in the process. The result of this action and counter-action of forces on each other may be the formation of compounds of elements and a state of equilibrium or diffusion of energy or both. The bulk of energy, manifested by the Absolute, is transformed at this stage into the corpuscles, electrons-protons etc., atoms, molecules of various constitutions and also into diffused energy in the form of heat, pressure etc. Now, due to further manifestation or individuation of energy there appears life in some compounds of elements which lie in an environment suitable for its sustenance. By suitable environment we mean the presence of such entities as are necessary for the maintenance of life which is to evolve later. We find such entities to be air, water, some carbon compounds, temperature and pressure in a particular range, and the absence of certain entities like alpha rays, etc. An elaborate preparation is therefore necessary for the emergence of life. The principles of change inherent in the very existence of entities like corpuscles of energy, electrons-protons etc., the chemical elements and their compounds are oriented by the Absolute in such a way that they ultimately create a suitable environment for the emergence of life.

Emergence of life cannot be explained as only an accidental product of changes in matter because a so-called accident, according to us, is never accidental. An event, may it be called an accident or anything else, must always have reasons behind it, even though the said reasons are unknown to us. And the reason according to us, behind the evolution of life is the principle of manifestation. The principle of manifestation differentiates and individuates energy through integration and so the stage after the evolution of elements and their compounds is the evolution of life. That the evolution of life is not accidental is further proved by the fact that an elaborate preparation was made before its emergence. It may here be said that life could have emerged even without any impelling force working behind it. Life appears only as a property of certain compounds of elements under certain environmental conditions and the same compounds and conditions act for its sustenance. We accept this explanation but in addition to it, like only to say that the emergence of life and its continuance are not due to an accident. These are only the results of a course determined by the principle of manifestation that works on energy.

Now, why life should appear only in some compounds of carbon and only in some environmental conditions and not in other compounds and in other environmental conditions is a riddle for us. It may be said that life emerges in some carbon compounds in some environmental conditions because the same are required for its sustenance. But then it could be so with other compounds and other environmental conditions also. However, it is a fact that the compounds of elements and the environmental conditions that evolve life are also required for its sustenance. It shows that an inorganic compound becomes organic and evolves life in it when it develops the technique of absorbing materials of its own kind or of dissimilar kinds and can convert the latter into materials of its own kind. The organic compound which is to manifest life is also required to maintain more or less the same collocation of energy in it. In so doing, the compound acquires a peculiar exclusiveness of its own and is individuated. Life appears in such an individuated compound because the conditions that evolve life are likely to prevail here and as such life may continue to exist. This, according to us, is the reason why life appears in organic compounds only. Life emerges not to be destroyed soon after its emergence. It emerges to fulfil some purposes of the Absolute. The purpose behind its manifestation, as we have said earlier, is to develop a freedom of will in it, and for this, it needs continuity in its existence.

As the will of an organic being is thwarted every moment by its environmental conditions and as there are wide gaps between its desires and their fulfilment, the said being develops an urge to continue to live and to acquire full freedom of its will. This is the root of the urge for procreation. A living being wants to fulfil his unfulfilled desires in the life of his offspring. To overcome obstacles that thwart his will, a living being takes lessons from his own experiences as also from the experiences of other living beings. Generations die out leaving lessons for generations who come later. Complete freedom of the will cannot be acquired by ordinary mortals. But with their limited freedom they pull on, determine to some extent the courses of movement of the universe and provide sports for the Absolute. Full freedom of the will, in the exact sense of the term, means that there is no gap between the desires of a being and their fulfilment. This freedom of the will thus means omnipotence and the being who has acquired such freedom of the will is obviously one with the Absolute.

Now, carbon compounds of some specific collocation under certain environmental conditions develop the peculiar capacity of absorbing materials of their own kind, or of dissimilar kinds to convert them into materials of their own kind, and also of maintaining more or less the same conditions in their structures. The simple fact is that other compounds do not develop this capacity. The said capacity of some carbon compounds is a concrete form of the principle of manifestation or individuation which means the development of a peculiar property of exclusiveness in these compounds. As this property of exclusiveness is present only in some compounds of carbon, life evolves only in them.

The field having been made ready for the oncoming of life and the possibility of its continued existence for further development, life evolves in some carbon compounds. As we have shown earlier, life is a form of consciousness which is present in energy of every form and collocation. We have also shown that energy implies consciousness in it and that consciousness also implies energy in it. Consciousness, in this sense, is present in the compounds of carbon which evolve life in them. When a carbon compound has individuated itself or has developed an exclusiveness of its own, we say that it has evolved life in it. The behaviour of the said compound of carbon which shows a tendency of absorbing materials from outside for conversion to materials of its own kind is due to the manifestation and work of consciousness. This kind of manifestation and work of consciousness is the principle of life. We have already shown that life implies also a mind in it which works as its central organ for the purpose of discrimination. Life and its activities cannot be chaotic, for in that case it will not be able to continue to exist. The very work of a carbon compound in absorbing materials from outside show the presence of a principle of discrimination in it because the said compound is to choose and absorb only those materials which are beneficial to it and to reject and avoid those materials which are harmful for it. This work of discrimination, even if it may be very feeble, is done by life through its central organ, called by us the mind. When the work of discrimination is very feeble, or even not perceptible to the organism itself, the activities of the said organism are said to be only instinctive. But as these instinctive activities also show an exercise of choice between beneficial and harmful results for the organism, the said activities are discriminatory in nature. Evolution of the principle of individuation and exclusiveness thus means the evolution of life and mind.

When life evolves in a carbon compound, the whole structure of the compound becomes a living creature. The whole structure becomes a living cell, because, at the earliest stage of life, the activities of the whole structure are confined to matters relating to its nutrition only. At this stage the whole structure of the organism is engaged in absorbing materials from outside so as to maintain more or less the same condition in it. The mono-cellular organism draws in materials from outside, absorbs a part of the said materials and expels the rest from its structure. The loss of energy due to changes in the structure of the organism is made good with the materials absorbed from outside and the requisite range of temperature is maintained so that the principle of life may continue to exist therein. In course of this metabolic process the living cell also grows in structure. A stage comes in course of changes when the cell is unable to absorb sufficient materials from outside to keep the torch of life in it burning. It is unable to convert materials absorbed from outside into materials of its own kind. The cell at that stage cannot repair its worn out body. There will automatically be a loss of temperature in the cell and the condition that prevailed in it when life was manifested, will be changed. This change will cause life to discontinue to exist and the organism will die. Being an entity having many limitations, the body of a mono-cellular organism cannot continue to function in the same way for ever because the said body is after all a collocation of energy only and as such is acted upon by other objects which change its structure. When the body fails to maintain the required condition, life in it ceases to exist. It may be said that the incapacity of an organism to maintain the required conditions in its body is the approach of death towards it.

At death the body of an organism does no longer maintain its composition and so a disintegration of the collocation of energy starts in it. The elements and compounds in the body now behave like inorganic objects and ultimately form compounds with other objects. And what about consciousness that appeared in the organism which now is dead? This consciousness, like the body of the organism, was not created out of nothing. The body of the organism was composed of energy and at death of the organism this energy was only transformed, and not lost because there is nothing beyond the Absolute to absorb it. Similar is the case with consciousness. Consciousness that appeared in the organism was not an entity created out of nothing. It was only a part of the universal consciousness or the consciousness of the Absolute. Unlike energy, consciousness cannot divide itself into its various elements because its constituents do always exist in an integrated form and as such the said elements cannot totally differentiate themselves from each other. Like energy, the elements of consciousness cannot be confined in separate corpuscles. If such confinement of consciousness could have been possible, consciousness could not have been integrated again, because there is no connecting link for the purpose. But in the case of energy, it is otherwise. The different elements of energy lie in an integrated form in their unmanifested state. When this energy comes to its manifested form through the principle of manifestation its various elements are separated one from another. In the manifested state these elements of energy do never integrate themselves again, for, that is the very meaning of manifestation. As there is no question of the integration of energy again, there is no necessity for a connecting link. But consciousness is an entity which exists always in the manifested state and as such the principle of manifestation or differentiation does not work on it. There being no connecting link in consciousness to unify its various elements, any re-integration of such elements would have been impossible, had they, somehow, been separated once. Division of consciousness into consciousness of different individuals is due only to the limitations of individuals, i.e. due to the limitations of their physical bodies. Individual consciousness is only consciousness, undeveloped and finite. Moreover, consciousness is such an entity that its elements do always exist in an integrated form. If they do not, the entity will cease to be consciousness at all.

At death of the organism, therefore, its life-mind complex or consciousness returns back to universal consciousness though it retains its individuated form. Individuated consciousness is universal consciousness as conditioned by the body of an organism. Now, consciousness, as released from a body due to death, has, through its experiences while in the living organism, reached a particular stage of its progress towards attainment of infinitude. As a principle for discrimination, this individuated consciousness worked for the well-being of the body and it had all along a sense of owning the body. It worked for the body as if it itself was the body. The element of will in this individuated consciousness willed and had desires and aspirations concerning matters relating to the body. After death of the organism, the released consciousness will therefore have a natural tendency to fulfil those desires and aspirations. This will to live in a body again, this thirst for bodily life or embodied existence, brings it down to material life again and attaches it to that very organism through which this individuated consciousness is likely to fulfil its desires and aspirations. This, according to us, is the cycle of birth and death of organisms.

The principle of manifestation that worked on the bulk of

energy, released from the Absolute, went on differentiating or individuating it and energy went on assuming different forms. Through the forms of corpuscles, electrons-protons etc., energy attained the state of organic compounds and then evolved life or consciousness in them. In order to keep the individuated form of elements intact, the principle of manifestation or individuation now acts as a principle of discrimination through consciousness. Individuation of an organic compound implies a discrimination between itself and other entities. In a sense, it may even be said that individuation is nothing but discrimination between the individuated entity and other entities and to discriminate is to individuate different entities. For an organism, progress towards its individuation means an increase of its work of discrimination. The principle of manifestation that implies differentiation and individuation thus also implies discrimination. There is a correspondence between the stages of individuation of an organism and its power to discriminate, or, in other words, there is a correspondence between the collocation of energy in an organism and its level of consciousness. Development towards the organization of energy in an organism means also the development of consciousness in it. As there is a link and correspondence between the stage of individuation of an organism and the stage of growth in its life-mind complex, the said complex, having been separated from an organism due to death, will seek for being attached to that organism due to be born which may suitably house it for further development in the individuation of the body and for further progress in the said life-mind complex. Thus, according to us, are death and birth of organisms linked with each other.

A mono-celiular organism starts its life with the sole function of drawing in materials from outside, absorbing a part of it and expelling the waste products. But the course of evolution makes this simple function of a mono-cellular organism a complicated and interlinked series of various activities. Evolution of energy up to the stage of carbon compounds was determined by the four principles, viz., the principle of change, the will of the Absolute, the principle of manifestation and the scientific laws specific for energy. But after the emergence of life-mind complex, or consciousness, in the said carbon compounds, the development of the body of an organism is determined to a certain extent by this complex also. Manifested energy is a constituent element of the Absolute and the consciousness implied in it is the consciousness of the Absolute. The difference between consciousness of the Absolute and consciousness of an organism lies in the fact that the former has no limitations in it whereas the latter has limitations imposed on it by its body which is limited in extent. Through the processes of discrimination, integration and assimilation of facts an organism can enlarge its consciousness to an infinite extent. Now, energy and consciousness being entities interlinked the changes in one will bring in changes in the other. The said changes will be radical or insignificant according as the changes in the other is radical or insignificant. It also means that a change in one forces the other to change itself. We have already said that energy, in any form, implies consciousness in it. So an organic compound, or even an inorganic compound, implies consciousness in it. But this consciousness may not have attained the level of a life-mind complex. There will not then be any influence of the said complex in the developments of such compounds. When, on the other hand, an organic compound evolves a life-mind complex in it, the said complex starts working on it as a principle of discrimination. This work of discrimination is an advanced sequence of the principle of manifestation and it is necessary for the individuation of an organic compound. To effect proper individuation of the organic compound, the life-mind complex starts functioning as a principle of discrimination and guides the organism in its activities and developments, and in course of this process it itself makes progress towards its attainment of infinitude. Thus, after emergence of life-mind complex in an organic compound the said complex turns out to be a strong force, though not the only one, in determining the course of developments in the body of that organism.

A mono-cellular organism draws in materials from outside and absorbs that part of the said materials which it can transform to a compound identical or very similar to the compound of its own body. So long as the organic compound had no life-mind complex in it, it had no mechanism of drawing in materials from outside. But after its emergence in it the organic compound starts behaving in a different way. Can its new behaviour be a result of any mechanical forces working on it? It cannot be so, in view of the fact that the environmental condition before and after emergence of life in the compound did remain the same. Therefore, the change in behaviour of the organic compound must be due to the work of life-mind complex on it. The work of this complex on the organic compound changes the whole pattern of movements in the said compound.

We have shown earlier that the function of a life-mind complex is to discriminate between objects and thereby to help the body, in which it has evolved, in its movements towards individuation. Now, the very first thing necessary for an organism is that it is to keep it alive. For this, the organism is required to maintain a particular condition in its body in respect of temperature, pressure and organization of elements and compounds. To maintain the required condition the organism draws in materials from outside for absorption. The processes involved in this regard help the organism to adjust the temperature, pressure and organization of its body elements. The life-mind complex of the organism discriminates between things and directs the organism to draw in suitable materials and to avoid other materials that may be harmful for it.

The principle of differentiation works on the body of an organism and its movements are guided by the life-mind complex through its work of discrimination. The body serves the purposes of the life-mind complex in it because all that it does, it does under the guidance of the said complex. The life-mind complex of an organism will determine what is beneficial and necessary for the body and will direct the body to move accordingly. The principle of differentiation that works on the body orients the changes in the body towards fulfilment of the directives of its life-mind complex. To justify this view we argue in the following manner. The very purpose of evolution of a life-mind complex is to evolve a principle that can guide the movements of a body towards its proper individuation. Due to manifestation, energy goes on differentiating itself and also integrating its like types till it reaches the level of formation of carbon and other compounds. These compounds are not properly individuated entities because they have no power of apprehension. The life-mind complex is therefore evolved in some carbon compounds to individuate them through the process of discrimination. Discrimination, which is the function of life-mind complex, is thus a process continuous with the process of differentiation. Consciousness, or the life-mind complex, discriminates, and discriminates for the purpose of guiding the organic body so that the said body may adopt certain courses of movement and may avoid certain others. In this sense, all the vital functions of the body are guided by the life-mind complex. Let us take an example. At a certain stage in the evolution of man a set of teeth are formed on his jaws. A man must first absorb materials necessary for the formation of teeth and then only he can grow teeth on his jaws. Calcium is the basic material for formation of teeth but this calcium is also the material necessary for the formation and growth of bones. After absorption of calcium the human body can utilize it for the formation and growth of bones instead of for the formation of teeth. Actually, both the bones and the teeth are grown by a human body. We chew our solid food with our teeth. It is not that we chew our solid food because we have teeth but that we grow our teeth first so that we may chew our solid food. Keeping thus in view our aim of chewing solid food, we grow our teeth. The purpose is determined first and then an attempt is made to fulfil that purpose. Now, determination of the purpose is a function of consciousness. The body of a human being, being only a collocation of energy, cannot therefore determine the purpose for the fulfilment of which teeth are to be grown. The body can grow teeth but only under the direction of its life-mind complex. The life-mind complex feels the necessity of teeth for the human beings who are to chew their solid food and directs the movements of the human bodies in such a manner that teeth are ultimately grown on their jaws. Again, why do the teeth grow on the jaws and not on any other place in the body? It is because the purpose of teeth is to chew the food and it cannot be done unless the teeth grow in the mouth. The body of a human being does not know where a quantity of calcium is to be sent and teeth are to be grown. If the body does know it, it does through the principle of life-mind complex or consciousness. It may here be asked: how could energy move towards the formation of carbon compounds before evolution of its guiding principle, the life-mind complex or consciousness? Our answer here is that before the evolution of consciousness, energy was guided in its movements by the principles of manifestation, differentiation and integration. Energy was undergoing changes through the processes of differentiation and integration and was moving towards its individuation but could not attain that stage because it was still under the control of the Absolute. It did not yet develop any principle of relative independence to carve its destiny and as such was yet to attain the stage of individuation. To complete the process of individuation therefore life-mind complex, or consciousness, evolved in carbon compounds. Consciousness completed the process through discrimination. The principles of differentiation and integration functioned as the guide in respect of the movements of energy up to the formation of elements and their compounds but thereafter the consciousness evolved also functioned as the guide in respect of further developments. The Absolute had full control over the developments of energy up to the formation of elements and their compounds but thereafter, due to individuation, the living beings also share the same controlling power with the Absolute. A living being, which is an individuated entity, cannot have any control over the workings of the principles of change, the will of the Absolute, manifestation or differentiation and scientific laws relevant to and inherent in energy. An individual cannot exercise his will over the workings of the said principles. He cannot arrest the change that his body undergoes every moment, he cannot have any control over the Absolute of which he is a tiny part, he cannot control or arrest workings of the principles of differentiation and integration that go on in the elements and compounds in his body and he cannot make or unmake the scientific laws that govern the said elements and compounds. But as an individuated entity he has independence in the exercise of his will over the activities in respect of his movements towards more individuation. In other words, as an individual he has control over his individuation. All activities for the sustenance of life and its ascent, through the works of discrimination, towards the Absolute are controlled by the individual himself. The formation and growth of the organs and limbs of a living being are also, to some extent, under the control of the said being.

In order to grow a muscle a gymnast conceives first a purpose that the muscle is to serve and then makes certain movements in his body for a period of time and ultimately a muscle is formed. When a man, short in stature, wants to be taller, he may undertake the practice of certain physical exercises and may ultimately grow taller by a few inches. If a man has a deformed limb, he may reform it by some physical exercises. All these examples show that a living being controls to a certain extent the formation and development of the limbs of his body. Now, these are all examples where a human being is conscious of what he does. He plans, he deliberates, he works and he modifies the formation of his body. These are all activities of the conscious region of a man's life-mind complex. But this conscious region is not the only constituent of such complex. There is also a subconscious and unconscious region in the said complex. Being constituents of the life-mind complex, the subconscious and unconscious region of mind also guide the activities of a body. As we do not come to know what is going there on inside the body we consider all changes in the body as only automatic developments of energy. But that they are not so is proved by the fact that morbidity in the mind of a man affects the functions of his body. The healing process in the body, if there be any, and the general growth of the body are hampered due to morbidity of the mind. Worries about a possible calamity affect the normal functions of one's body, even if the said worries are not allowed to dominate the conscious region of one's mind. The urge for life (or, vitality) makes one live and grow till that urge is overcome by any other force. This urge can seldom be analysed by us because it lies in the unconscious region of the mind. The vitality or urge for life heals up many of the minor wounds in a living body even if no medicines are applied. All these examples show that the life-mind complex takes part in many of the functions of a living body. As at the appearance of life-mind complex in a compound of elements, the said compound moves in the way we call the vital functions, it may be said that the said complex has a close connection with the vital functions. Being a principle of consciousness, life-mind complex cannot be the movements of energy, called the vital functions. The only other way in which the life-mind complex can have any relation with the movements of energy is by influencing and guiding the latter. The vital functions are movements of energy but these movements are not only guided by the scientific laws but by the said laws as modified by the inherent laws of life-mind complex or consciousness. As the life-mind complex of a living being controls the growth and development of some of his limbs and organs we find no reason to believe that the said complex has no role in the growth and development of organs like the stomach, the heart, the lungs etc. and even the limbs like hands, feet etc. In accordance with our analysis of the life-mind complex, as done in a foregone chapter, we may say that the said complex feels the necessity of those organs and limbs for the living being concerned and directs and guides the movements of energy in his body towards the formation and development of those organs and limbs.

The life-mind complex of a living being is evolved in a compound of elements and it is therefore related to the latter. The welfare of the body depends on the welfare of the mind and vice versa. The function of the body is to grow and develop through changes and the function of the mind is to discriminate between things and integrate its knowledge for the purposes of the body for which it has evolved in it. We have said earlier that a particular organ is grown and developed in the body for a definite purpose to be served by it. An organ is grown first and the purpose is served thereafter. The determination of purpose is a function of consciousness and as such it cannot be done by the body which is only a collocation of energy. The normal function of the mind is to deliberate for the purposes of the body. The mind is to determine what organs and what limbs are to be grown in the body and it is to direct the movements of energy in the body accordingly towards the formation of those organs and limbs. The normal function of the body is to act in accordance with the directions from the mind. If there is any discord between the actions of the body and of the mind, both the body and the mind will be out of order and as a result there will be an abnormal death of the organism, with which we are not concerned at present.

Life is manifested in a compound of elements and that is done not definitely for the purpose of its annihilation. Life cannot appear in an organic compound only to disappear from it. Evolution of life also means its continuance. An organic compound evolves life to further its own manifestation and as such one of its basic functions is to sustain that life in it.

Being only a collocation of energy, the organic compound cannot determine what movements are conducive to the continuance of life in it. Life-mind complex therefore comes forward to direct and guide the movements of the body for its own continuance in it. The body does not know what movements may force it to move towards death and what movements may be beneficial to it. It therefore follows the directives from life-mind complex and acts accordingly. It flies away from the places of possible danger and hurries to move towards things beneficial to it. In the case of movements of its outer limbs, the body follows the directions of the mind. The said limbs are governed not only by the mechanical laws of energy but also by the laws of mind. They are governed by the mechanical laws in the sense that the processes of disintegration of existing compounds and formation of new compounds go on in them. The movements of energy in the limbs are thus governed by the principles of differentiation and integration as it used to be before the evolution of a life-mind complex in it. The limbs are governed by the laws of mind in the sense that the outer movements of limbs are governed by the mind. It may therefore be said that the movements of limbs are not only governed by the mechanical laws of energy but also by the mind. As the mind itself is not the energy we are speaking about, it can only direct and guide the differentiation and integration of energy towards the formation of limbs for some purposes and can make the said limbs move in some definite ways to fulfil those purposes. Now, if such be the role of life-mind complex in the growth and movements of outer limbs, why should not the said complex have any role in the growth and movements of the internal organs of an organism? The internal organs of a living being are of the same nature as those of his outer organs, the only difference being that the internal organs have a more vital role in the life of an organism.

As the body of an organism has its outer and inner organs, its more vital and less vital organs, so also the life-mind complex has its outer and inner aspects, its more vital and less vital regions. As a living being can continue to live even if its outer limbs remain inactive, so also a life-mind complex continues to exist in the body even when its outer region ceases to act. Now, both the inner and the outer regions of a mind have their specific roles in the movements of energy in a body. As the outer region, or the so-called conscious region, of a mind guides only the movements of outer limbs and not the inner organs, so we may presume that the inner regions, or the so-called subconscious and unconscious regions, of a mind guide the movements of energy in the inner organs of the body, though there may not be a strict division of labour in this respect. Thus, according to us, the outer region of a mind guides mainly the movements of outer limbs of a body. The so-called subconscious region of a mind guides the movements of outer limbs in some cases, as in the cases of night-walking etc. and also guides the movements of energy in the inner organs. The inner most region, or the so-called unconscious region, of a mind works deeper into the very collocation of energy in the system towards further manifestation and individuation of the body. We know very little about this region of the mind. Most of its laws and workings on the collocation of energy are unknown to us. We therefore think that the growth and development of organs and limbs in a body are only automatic developments of energy. But, being purposive the growth and development of organs and limbs cannot be planned by anything other than a principle of consciousness.

The principle of differentiation which worked up to the stage of formation of organic compounds takes a new turn due to the appearance of life-mind complex in the said compound. To fulfil the requirements of life the organic compound now starts undergoing change in a different way and the principle of differentiation is modified by the life-mind complex. The process of differentiation goes on in the organic compound but now it works under the guidance of the life-mind complex which is a principle of discrimination. The principle of discrimination is concerned not with energy, as it is, but with energy as it works. It brings in a process of discrimination in the works of energy so as to orient such works towards individuation of the body and the mind concerned. Now, the main function of an organic compound after appearance of life in it is to maintain a particular condition in it so that life may continue to exist there. And what it means is that the organic compound is to draw in materials from outside, convert a part of them to compounds of its own type and to absorb them so as to make good the loss of its body due to dissipation of energy effected by the courses of change, as also to keep the temperature of the body within a specific range. This nutritional function is very simple in the primitive organisms. Life starts functioning in a primitive organism by directing it to absorb materials from outside and to convert a part of them to compounds of its own type. Before the appearance of life in it, an organic compound does not behave in the said manner. Life causes it to move in the way of absorption of materials from outside, conversion of some of those materials to materials of its own type and assimilation of the converted materials. This shows that life may work on the very core of energy and modify the latter's behaviour radically. It also shows that life-mind complex or the principle of discrimination is not less effective than the principle of differentiation on the movements of energy. The principle of differentiation, coupled with the principle of discrimination (due to the appearance of life), makes an organic compound behave in the clearer way of fulfiling a purpose. An organic compound manifests life for further individuation and as such one of its basic functions is to keep the torch of life in it burning. It therefore draws in materials from outside and converts it to compounds of its own type, for that is the basic requirement for existence of life in it. The said conversion of materials is nothing but a reorganisation of the units of energy in them. The function of reorganising energy is not anything new to an organic compound because the compound had this function when only the principle of manifestation worked on it. After the appearance of life now, in addition to the principle of differentiation, the principle of discrimination also works on the compound and causes it to reorganise the units of energy in a way conducive to the existence of life in it. The organic compounds in a living body undergoes change in a different way due to the workings of its life-mind complex on them.

We have said earlier that the life-mind complex is mainly concerned with the functions of energy in the body of an organism. When therefore the life-mind complex works on the nutritional functions of the body, i.e., its functions of absorption, conversion and assimilation of materials drawn from outside and expulsion of waste products, new organs are formed in the body. To meet the difficulties due to possible irregularity in the supply of food, the organism should have an organ for storage of food. For proper assimilation of food an elaborate system of digestive organs are necessary. For proper utilisation of gases from air, respiratory organs are necessary. For the distribution of assimilated materials to different parts of the body, a circulatory system is necessary. To collect various types of food for the growth of various organs, organs for locomotion are also necessary. The life-mind complex goes on differentiating the simple process of nutrition into an elaborate system necessitating thereby the formation and growth of various organs for the purpose. An organism requires sense organs for the purpose of sensing objects in the surroundings. Thus a blurred awareness of objects in the surroundings is differentiated by the life-mind complex into its various elements for the sensing of which various organs are formed in the body. For progress in its function of discrimination, the life-mind complex requires more individuation of the body to which it is attached. So, for its own purpose of discrimination, the life-mind complex guides energy in the body towards formation of an elaborate system of organs for digestion, blood circulation, excretion, respiration, reproduction, locomotion and sensation. The said organs are formed in the suitable places of the body, e.g. the set of teeth are formed on the jaws, the stomach in the cavity of the body, the lungs and heart in the most protected area of the cavity and so on. These organs are located in the places most suitable for their proper functioning. If it would have been a fact that the life-mind complex played no part in the formation of organs in a body, if the formation of these organs was due only to the workings of mechanical forces involved in the process, the stomach could have been formed in the head or the brain in the cavity of the body. There are no evidences to make us believe that the stomach was once formed in the head and then had been shifted to the body cavity for suitability of its functioning. Actual evidences show that the formation of an organ starts in that very place of the body from where it can in future function properly. There is no principle in bare energy to let it know from where an organ, to be formed by it, can function properly. A principle of consciousness must therefore work behind the formation of an organ. According to us, the life-mind complex, evolved in the body, is the said principle of consciousness that works behind the formation of organs in the body. The life-mind complex differentiates the simple function of nutrition in the body into an elaborate system of interconnected functioning necessary for further individuation of the body and also for progress in the process of discrimination going on in the said complex. The life-mind complex deliberates and feels the necessity of a group of organs in the body for the purposes of sensory, motor and nutritional matters. As a directing principle, the said complex guides the energy constituting the compounds of elements in the body towards the formation of various organs and limbs.

We are now in a position to show in outline the actual course of evolution of the material world including the organisms that live and die in it. The bulk of energy manifested by the Absolute goes on differentiating and integrating itself in the way we have shown earlier till it reaches the state of compounds of elements. All these changes in energy and all its developments are, according to us, purposive in the sense that the Absolute did effect them voluntarily. The Absolute manifested a bulk of its energy for the purpose of creating individuated beings with freedom of their will who could carve their own future. The Absolute selected some carbon compounds for the said purpose keeping the other compounds only subservient to them. Due to further developments in it, the principle of differentiation turned into the principles of discrimination and individuation and as a result there appeared life-mind complex in some carbon compounds. Now, after the appearance of life in it a carbon compound becomes individuated in the sense that it now comes under the influence of that life-mind complex. The life-mind complex has appeared in the body to use it as an instrument for its purposes of discrimination and as such it directs the movements of energy in the body in that line. And the body evolves life in it for the purpose of having a guiding principle in its movements so as to be individuated. The function of the body is to comply with the requirements of the life-mind complex in it.

The principles of differentiation and integration work on the energy contained in the body of an organism. The body of an organism in this respect is energy only like energy in other inorganic and non-living organic compounds. The difference between a non-living and a living compound is that while in the former, only the principles of differentiation and integration are in action, in the latter, not only the principles of differentiation and integration but also the principle of individuation is in action. Due to the presence of life-mind complex in it, a living body cannot integrate or disintegrate like non-living compounds. Whatever is done by a living body, is done under the guidance of the life-mind complex in it. The life-mind complex has twofold functions. It has evolved in an organic compound for the purpose of individuating it and also for the purpose of guiding the movements of energy contained therein. As a guiding principle, the life-mind complex has to look after the welfare and development of the body it is attached to. It has to direct the movements of energy in the body towards the formation of organs and limbs necessary for the purpose of its nutritional functions. It has also to guide these organs and limbs in their activities in connection with the sustenance of the body and its life. Secondly, as a principle of discrimination and deliberation, the life-mind complex has specific functions of its own. In course of its work of discrimination, the life-mind complex goes on acquiring knowledge of more and more entities of the universe and ultimately becomes one with the Absolute, which is infinite consciousness. As the ultimate aim of the life-mind complex is to be one with the Absolute, its work as a guiding principle of the body is also oriented towards that aim. The body is practically used by the life-mind complex as its vehicle for movement towards the Absolute. The movements of energy in the body are guided by the said complex in such a way that the body may help this complex in its work of discrimination. But whatever help the body may render to the life-mind complex, the body, being only a cluster of energy, has its own limitations. The body cannot maintain the required condition in it for sustenance of life for ever. In course of its undergoing changes every moment, the body ultimately reaches a stage when it fails to maintain the required collocation of energy in it and is disintegrated. Moreover, as a material entity, the body can help the life-mind complex to come in touch with the material world and to acquire knowledge of that world only. In course of its development and progress, therefore, the life-mind complex reaches a stage when the body does no longer act as an aid to its work of discrimination. At this stage the mind goes on discriminating and integrating its mass of knowledge unaided by anything material.

Appearance of life in a carbon compound makes it individuated. The said compound acts as the body of an organism and is under the direct influence of its life-mind complex. For its own existence, the life-mind complex requires the body to maintain the conditions necessary for the continuance of life in it. The conditions necessary for continued existence of life in a body includes a particular collocation of energy and a particular range of temperature and pressure in it. Due to changes to be undergone by the body every moment, the collocation of energy in it cannot remain the same. The body expends energy in course of changes and as such needs an intake of materials from outside so as to make good the loss of energy as also to maintain the required range of temperature and pressure in the body. This requirement of materials from outside is the urge of a living being for food. The urge for food is a mental phenomenon which causes the body to move in a particular way. Due to this urge the body draws in materials from outside, transforms the collocation of energy in them and assimilates them to make good the loss of energy in the body due to change. Before the appearance of life in it, an organic compound does not collect food for assimilation. Why then it does so after the apperance of life in it? It does so because life has now made it an individuated entity, which means that the compound is now a living body separated from every other entity of the universe. As a sharply defined area of the influence of life-mind complex in it, the body is required to maintain its integrity. It is to maintain its existence as an entity separated from every other entity and it also is to fulfil the requirements of the said complex. Due to the presence of a life-mind complex in it, the organic compound is not only a differentiated entity but is also an individuated being. In order to maintain its existence as an individuated being, the organic compound changes its behaviour radically and goes on absorbing food to keep intact the collocation of energy in it.

A living body intakes food, converts them to compounds identical with materials in it, and assimilates them so as to keep the collocation of energy in it intact. In course of this process a living body not only makes good its loss of energy due to changes but also grows in volume. The principle of individuation not only causes the body to maintain its individuality but also impels it to divide itself into two or more parts so that each such part may continue to exist in the form of an individuated being. This is the process of reproduction in mono-cellular organisms.

The body of an organism grows in spite of the hurdles it passes through. A stage however comes when the body can no longer adjust itself with the environmental hurdles such as shortage of food, sudden fall or rise in temperature and pressure or accidents. Due also to age the organs may lose their capabilities of collecting and assimilating food. As a consequence the body fails to make up its loss of energy due to changes involved in its very existence and goes on losing strength. A stage ultimately comes when the body completely fails to maintain its organic existence and it dies. Its life-mind principle departs from the body and waits for being embodied in a suitable organism again.

Now, the life-mind complex of the mono-cellular body in question not only guides the body but also deliberates and discriminates during its embodied existence. It tries to find out means for the body to cross over the hurdles in the environmental situations. It feels the necessity of an organ in the body for storage of food to overcome the danger of periodical shortage of food. At its direction therefore the energy in the body moves in such a way that after a long course of changes through generations the body ultimately grows an organ for storage of food. For proper assimilation of food different chemical substances are necessary. At the direction of its life-mind complex the body orients its energy towards the development of a system for the said purpose. Due to differences in quality of materials consumed by different organisms there will be minute differences in their body structure and the composition of elements therein, though as living beings they all will have the same or similar collocation of energy in their bodies. Due also to the aforesaid reason will there develop differences in the mechanism of consuming materials as food and their assimilation in the bodies of different organisms. The specific nature of organs and mechanism for nutrition in an organism are inherited by its offspring. The offspring retains those traits, inherited from its parents, which facilitate its mechanism of nutrition. It develops some more traits due to differences in quality of its food and also due to the workings of its life-mind complex. As a parent, it also transmits its specific traits to its offspring and this process of developing differences in structure and mechanism of nutrition goes on through the generations. Due to these differences an organism may look quite different from its ancestors.

For food an organism may either consume inorganic substances and convert them to organic compounds for assimilation or it may consume organic substances, if such substances are available, or it may consume both inorganic and organic substances. An organism takes up the first course if only inorganic substances are available in its vicinity or if it prefers to consume inorganic substances only even though both inorganic and organic substances are available near it. An organism's preference for inorganic substances may be due to its inability to assimilate organic substances or due to its adherence to the old habits only. An organism takes up the second course, as aforesaid, due either to the availability of only organic substances in its vicinity or due to its inability to assimilate inorganic substances or due to its preference for organic substances because of their easy assimilation in the body. The processes adopted by organisms for collection and assimilation of food are generally adopted by their offspring due to their inheritance of inherent traits from the parents. In course of living their lives through the generations, the organisms demarcate themselves into two groups—one group that develops a mechanism for assimilation of inorganic food and the other that develops a mechanism for assimilation of organic food only. Organisms of the first group are the plants and those of the second group the animals. There may also be a third group of organisms who are able to assimilate both inorganic and organic substances.

For proper assimilation of food, the life-mind complex in the organism directs energy in the body to move towards the growth and development of an elaborate system of organs. The life-mind complex does so because after deliberations it feels the necessity of such organs for the organism which is to live longer for further developments. The growth and development of organs depend on the nature of materials present in the body of the organism, the nature of stuff it draws in for food, the mechanism necessary for proper assimilation of the food and also on the traits that an organism inherits from its parents. For example, the plants who are to collect their food from the soil and air grow roots and leaves for the purpose but do not grow any organs for locomotion, which are unnecessary for them. But an animal, on the other hand, collects its food from different places and as such grows no roots but limbs for locomotion for the purpose. An animal inhales air not for collecting carbon from it but for collecting oxygen and so it grows no leaves but grows inhalatory organs instead for the purpose. As the original stuff in the body of an organism is to serve different purposes in connection with the growth, nutrition and development of the organism, so the mono-cellular stuff of the body gets differentiated, in due course, into a multi-cellular form and different organs are grown with them. The life-mind complex of an organism requires the organism to continue to live for ever so that it may make progress in its work of discrimination and may acquire more knowledge of the universe. The said complex therefore deliberates and feels the necessity of proper nutrition and growth of the organism. And this necessity is thus the main factor that determines the growth of organs and limbs in an organism. The organs necessary for effecting proper growth of a plant are not the same as the organs necessary for effecting growth in an animal.

The mono-cellular body of an organism gets differentiated into a multi-cellular form in order to serve various purposes of the body and the body also grows in volume due to assimilation of food. The bulk of the body gets increased creating a necessity of transmission of fresh food to different areas of the body. A circulatory system is therefore grown in the body. Systems in this way for digestion, excretion, reproduction, respiration, locomotion etc. grow and develop in the body of an organism. The growth of organs and limbs in an organism is determined by an organism's necessity for them, which again depends on the local conditions that an organism is to face. The exact line of growth of the organs and limbs thus differ from place to place and from one species of organism to another. The organs are therefore shaped differently in plants and animals, differently in different species of plants and in different local conditions, differently in different species of animals and differently in different areas of their habitation.

The abnormal growth in organisms and their deformities are, according to us, due to shortage of food, consumption of food in excess, poor or rotten quality of food consumed, improper assimilation of food and improper functioning of organs and limbs. It may also be due to an organism's defiance of the inner urges and instincts or even due to its movements under direction based on a misdeliberation and miscalculation of its life-mind complex. The quality and quantity of food consumed are two very important factors in the growth and development of an organism. The functioning of organs and limbs are also factors that determine the growth of an organism. According to us, the inner urges and instincts in an organism are the promptings of its life-mind complex. The inner urges and instincts disclose to an organism the inner workings of its life-mind complex. The specific function of life-mind complex, i.e., its work of discrimination and deliberation, depends on the existence and proper functioning of the organism and as such the inner urges and instincts, being promptings of the life-mind complex, cannot be anti-life instigations for the organism and these promptings cannot have any malevolent design behind them. The urges and instincts prompt an organism to do what is beneficial to it, what may let it continue to live and live longer. But sometimes an organism pays no heed to its inner urges and instincts and faces calamity. Sometimes, the urges and instincts are misinterpreted and misunderstood by the organism as a result of which the said organism may also have to face calamity.

The instincts and urges of an organism are related to the deeper area of its life-mind complex, or its so-called unconscious mind, and as such they reveal the inner workings of the said complex. The deeper areas of a living being's mind are not within the control of its will because deliberations and calculations are not done therein consciously. A living being exercises freedom of its will only in the areas of mind where it calculates and deliberates consciously, though the results of these calculations and deliberations may result in a clear defiance of its inner urges and instincts.

The inner workings and movements of energy in the body of an organism are determined by the deeper areas of its life-mind complex. An abnormal growth of the body of an organism is due to the movements of its life-mind complex on the wrong track. The function of life-mind complex is to individuate an organic compound and to guide it to grow in such a way that it becomes an instrument for the said complex which is to acquire knowledge of the universe and to enlarge its consciousness. It is also the function of life-mind complex to guide the organism in its movement for continuous existence as an individuated entity. As a principle of consciousness, the life-mind complex has its own line of movement and it is distinctly different from the line of movement of energy in the organism. But if instead the life-mind complex abandons its natural course of movement and entangles itself too much with the affairs of the body, then it is sure to fail in its duty. Also, as a guide for the movements of energy in the body, the life-mind complex, on such a wrong track, is sure to fail. The promptings of such a life-mind complex, in the form of inner urges and instincts, will therefore only misguide the body and bring peril to it.

Now, what form an organic body may ultimately assume in course of evolution? To give an answer to this question we must start our analysis from the side of life-mind complex because all developments in the body are determined by it. The life-mind complex is a principle of consciousness implied by the energy that constitutes the body of an organism. It remains implicit in the energy so long as there is no necessity for energy to be guided by it. Individuation of some collocations of energy, when so willed by the Absolute, makes it necessary for a part of the implicit consciousness in them to become explicit and as a result, a life-mind complex is evolved in them. Being a principle of consciousness, the life-mind complex has the potentiality of reflecting the whole universe, but being manifested in energy contained in a body that has its limitations, the life-mind complex will also have some limitations. To overcome these limitations, the life-mind complex is to cross the barrier of the body, not by defying it forcibly, for that only shuts out the possibility of its acquiring knowledge through the work of discrimination, but by using the body as an instrument for its act of discrimination so that it may acquire knowledge and enlarge its consciousness and may overcome the limitation of the body in due course. The life-mind complex therefore goes on individuating the body and in due course develops in it sense organs for the purposes of the body and also for its own purpose. The sense organs are to serve the body by collecting information about the environment for nutritional and other purposes. The sense organs are also to serve the life-mind complex by acting as instruments for its work of discrimination and integration of knowledge, enlarging thereby the sphere of consciousness in the said complex.

At the earliest stage of evolution of an organism, its body is mono-cellular and it has only a vague and blurred sense about its environment. The whole body at this stage acts as an organ for a vague cutaneous sensation only. The organism cannot differentiate the different elements of such vague sensations. In course of its progress towards individuation, the different parts of the body of such an organism are differentiated and distinct organs of sense are grown in the body. The organs of sense collect information about the environment and transmit the same to the brain which acts as the central organ for storage and co-ordination. The life-mind complex analyses the information stored up in the brain (or, in any such central organ), deliberates on them and sends its messages to different parts of the body through the brain and the nerves. Now, the sense organs are not only to serve the purposes of the body but also of the life-mind complex. If the body has no need for a particular organ of sense, will it not grow the said organ? It will, because the life-mind complex may have need for it. An organ or limb may serve the purposes of both the body and the life-mind complex. But it does not mean that the said organ or limb must serve the purposes of both the body and the life-mind complex. It may serve the purposes of either the body or the life-mind complex.

We are now in a position to say something categorically about the forms that organisms may assume in course of evolution. The actual course of development that an organism will undergo depends on the environment in which it is placed. But it may be said that an organism at an advanced stage in the scale of evolution is likely to enjoy a very long life and it is likely to have a group of very sensitive and powerful sense organs including a very powerful brain. The five or less typical sense organs that we find in the animals on our earth may not be the only possible items of such organs. If the environmental conditions are changed, then, for its own work of discrimination and integration, the life-mind complex may require a different number and type of sense organs and a different type of brain. Till now, we find human beings to be the most advanced form of a living being in the scale of evolution on earth. But it does not mean that human beings will live on this earth forever. If the human form does no longer remain suitable for its work of discrimination and integration of knowledge, the life-mind complex may replace this form by another. For its own purposes of development and progress, the life-mind complex requires a healthy body that will exercise its developed sense organs properly and will continue to live for a long time. The said complex therefore directs energy in the body to grow and develop in that line. But if an organism defies or obstructs this inner impulse of life and exercises its will in a different way, then the body will assume a form that will not serve the purposes of the life-mind complex. In such a case, therefore, the said complex will have to replace this form of body by another suitable for its purposes.

Chapter 6
Activities of Men—Economic, Artistic, Philosophic and Religious

The first principle behind this universe, as per our analysis, is an infinite consciousness implying an infinite energy in it and we prefer to call this entity the Absolute. It is the only substance in the true sense of the term and there is nothing beyond it. There can be nothing outside it and nothing other than it. All entities, material or spiritual, pertaining to matter or pertaining to consciousness, are only modifications of the Absolute. The Absolute undergoes change every moment, for change, according to us, is the very essence of existence. The infinite consciousness of the Absolute is ever explicit, because, the very nature of consciousness is to be explicit. Consciousness may of course remain unmanifested if there exists a barrier to keep it clouded. But in the case of infinite consciousness of the Absolute there is no such barrier. The infinite energy of the Absolute cannot limit the said consciousness because the latter implies the existence of the former. The infinite energy of the Absolute may, however, remain implicit and it actually does so. For its very existence the Absolute must have energy in it in an implicit form. Energy may thus be both explicit and implicit in form.

Now, for the Absolute, manifestation of energy, except for the purpose of its own existence, is not at all necessary because there being nothing outside the Absolute and nothing unknown to it, it does not matter whether its energy is manifested or not. There being nothing outside it, the Absolute cannot have a feeling of want for anything because such a feeling necessarily means the existence of things outside it. The Absolute continues to exist, and may exist for ever. Every entity that exists or may exist, every form of entities, actual or potential, is known to the Absolute. Nothing being unknown to the Absolute, the Absolute enjoys no novelties in situations. From our human standpoint we find the condition of the Absolute as that of a monotonous existence only. The Absolute may weave out the universe by combining its fibres, actual and potential, in any manner it likes and may bring varieties out into existence. But this also cannot create novelty for the Absolute because the said varieties are not unknown to it. It cannot, of course, be said that the Absolute is bound to feel the monotony of its existence or must try to overcome that monotony. If, however, the Absolute feels the said monotony, in the human sense of the term, it will have to create a principle that may exhibit novel situations before it. As the Absolute may do so, though not necessarily, we may say that it is only an activity of sport for it. If the Absolute desires to overcome the said monotony it evolves the world so as to witness novel situations. We, therefore, propose to explain the evolution of the world from this point of view.

For creation of entities in its own image and to witness novel situations, the Absolute must bring out a principle of freedom of the will in those entities, limited though that freedom may be. The creation of such a principle is necessary because without some independence in its movements an entity cannot display something not previously known to the Absolute. With the said end in view the Absolute manifests a bulk of its implicit energy and due to works of the principle of manifestation the said energy undergoes changes in a particular line and in due course the corpuscles of energy, electron-protons etc., the chemical elements and their various compounds are formed. Evolution of all these entities is only the preparation of a ground for the appearance of entities that may act as agents for creating novel situations. In this sense it may be said that even the stars and planets, the sky and the seas, land and water, temperature and pressure were all made to appear for the sake of those entities or creatures.

Due to manifestation of energy and its integration, first the corpuscles and then the fiery balls of it, the stars, came into existence. Due to radiation of heat energy some stars got disintegrated and in course of time planets and such other objects came into existence. Due also to radiation of energy, electrons, protons, atoms, molecules and the chemical elements were formed. Then were formed the compounds of elements and in some areas of the material universe water and atmosphere came into existence. Due to further radiation of energy the atmospheric pressure and temperature ultimately reached a level suitable for the oncoming of creatures. The general principle of manifestation ultimately got a channel to develop further into the principle of individuation in some compounds of carbon and consciousness implied in the energy in those carbon compounds manifested itself in the form of life. An entity is thus formed which has both energy and consciousness in the manifested form and has somehow developed a principle of relative independence because of the freedom in its will.

The aim of the Absolute is to bring out creatures that may have independence in the exercise of their will. So, the principle of manifestation cannot stop functioning after the formation of chemical elements and their compounds only. It now comes to a higher stage of its application and gets transformed into the principle and process of individuation. The principle of individuation is an advanced form of the principle of manifestation. We have shown earlier that manifestation is nothing but differentiation of an entity from other entities, and it also implies integration of the differentiated entities. When an entity is manifested, it comes from an implicit to an explicit form. An entity in its implicit form had no distinguishable identity of its own. It had no boundary to confine itself in and as such, it was not distinguishable from other entities in the integrated mass. When the entity comes to its explicit form it gets a boundary to confine itself in. Prior to its manifestation it was present everywhere in the integrated mass of implicit elements but due to manifestation it is now present only somewhere in the said mass. As due to its having a boundary of its own the entity assumes a new form, the entity is separated from the integrated mass, or, rather, differentiated from it. Manifestation thus means differentiation of an entity. And when an entity is thus differentiated and it gets a boundary of its own its different elements are confined in it. The various elements of the entity now come closer to each other and due to their confinement within a definite boundary a closer and compact relation grows amongst them. The entity now displays its attributes in a clearly defined way. All these developments only mean an integration of various elements of the entity. Manifestation of an entity thus means its differentiation from other entities and the integration of its differentiated form. But there is no principle inherent in a differentiated entity to keep the integrity of that entity intact. Due to impact of other entities on it, the integrated form of the entity may get lost. Differentiation and integration cannot therefore be the last stage of manifestation of an entity. Manifestation, in the full sense of the term, must therefore include a principle to keep the integrity of a manifested entity intact. That principle, according to us, is the principle of individuation or a sense of owning the integrated entity and a sense of responsibility to keep the entity intact. Individuation thus is a form of consciousness. So, for further advancement of the principle of manifestation, consciousness must appear in a differentiated entity. The sense of owning an entity (defined by a clear-cut boundary) and the sense of identifying the consciousness involved with the entity is the principle of individuation, according to us. When the principle of manifestation reaches the stage of individuation in an entity, the conscious entity excludes itself from all other entities and develops an ego-sense. It develops a tendency to think and act with a view to safeguard the interests of its own body and mind. An individuated entity begins to look after its own interests and tries to keep its integrity intact. It resists or tries to resist all forces that threaten to disintegrate its being. It discriminates itself from others and tries to continue to live as an individuated entity. This is the lust for life.

The developments of energy up to the formation of carbon compounds were all determined by the Absolute through his will, the principle of change, the principle of manifestation and the scientific laws. But now the ego-sense, generated due to individuation of an entity, also becomes a determining factor for the movements of that entity. The sense of owning the body, the sense of responsibility for the welfare of the body, or consciousness, also now works along with the four principles mentioned above in determining the course of development of the individuated entity. This egoism, developed in an entity, implies a will in it to choose one out of two or many courses of its action. In the case of inanimate entities, only the four principles mentioned above, or these four forces, determine the actual movement of the entity. The resultant of these four forces impels the inanimate entity towards that movement which is the only possible one, there being no other alternative movement left for it. When a stone falls from a hill-top, its movement downwards is the resultant of the four forces working on it and the stone itself has no urge to look after its own welfare nor can it resist the movement. The course of movement of a falling stone is fixed mathematically. But in the case of an animate being the situation is otherwise. An animate being has a mind and a will of its own. It has an ego-sense and it tries to look after its own welfare. A living being tries always to keep its integrity intact. Whenever a situation arises posing any threat to its integrity, whenever there arises the question of greater or lesser welfare of its being, whenever there arises the question of greater or lesser harm, the said being exercises its will to choose one out of two or many alternative courses of its action. The ego-sense thus implies an exercise of the individual will. A stone does not look after its own welfare but a living being does. A stone has no principle of will in it but a living being has. The principle of will present in a living being also implies an exercise of that will. A living being exercises its will and determines one out of two or more courses of its action. It cannot of course determine all movements of its body and mind and it is so because the individual will is operative in a limited sphere of activities only. An individual can choose to go by this road or by that road but it cannot stop circulation of blood in its own veins. There are some movements in its body and mind where only the four principles—the will of the Absolute, the principle of change, the principle of manifestation and the scientific laws—are operative and its individual will plays no part. The exercise of one's will in determining the course of activities in one's life implies a freedom in that will. Unless there is freedom in the will, how can it be exercised? Here at this stage of the evolution of energy therefore the Absolute fulfils his purpose of bringing out creatures who may create situations not foreknown to him.

After the emergence of freedom in the will of individual creatures, creation in the real sense of the term begins. An individual faces new situations every moment. An individual does not know whom he will meet on the road when he goes out in the evening for a stroll. He does not know before entering a forest what flowers are going to present their blooming faces before him. He does not also know how he will react while meeting people on the road or blooming flowers in the forest. The exact patterns of his behaviour in unknown situations are not foreknown to the individual concerned. The patterns of behaviour determined by an individual are therefore his creations. These are creations not in the sense that they have come out of nothing but in the sense that they were new and not foreknown to him before their actual occurrence.

Now, the situation viewed at from the standpoint of the Absolute is this: All that actually happened in the universe are known to the Absolute, because he is an omniscient being. All that may happen in the inanimate world are also known to the Absolute. The inanimate world has in it no principle of freedom of the will and as such it is within the full control of the Absolute. There being no individual will in that world, the question of its exercise to carve out patterns of movement not foreknown to the Absolute does not arise. But the works of individuals done by exercising their free will are not foreknown to the Absolute. The exact patterns of change that the animate world is going to affect cannot be fully determined by the Absolute. The Absolute knows the possible courses of change the animate world may undergo but which one out of many alternative courses will be actualized is unknown to him. The individuals who have been given freedom to exercise their will freely will do so and will choose their courses of action freely and the Absolute cannot determine their activities. It may here be said that the particular course of action an individual is going to choose and the changes the animate world will undergo are only the explicit forms of what were implicit in their antecedent states and as such these changes cannot be said as creations of individuals. Our explanation here will be this: the course of action actually chosen by an individual and the changes brought thereby were implicit in the antecedent states but these were implicit along with other alternative courses of action and possibilities of change. So the exercise of individual will makes only some changes explicit keeping others to remain implicit forever. The individual will thus determine the change the world actually undergoes. As the individuals will is exercised by the individual himself and not by the Absolute, the latter cannot determine the changes in question. It may still be said that the individual will, being an entity produced by the Absolute, cannot have any freedom in it. The freedom that an individual feels to have is only a semblance of it because he himself, including his will and all, is an entity produced by the Absolute. Our explanation here will be: the very principle of individual will means that it has an autonomy in it. An individual feels to have it and, according to us, the very purpose of the Absolute in bringing the individuals out into existence is to give them freedom to exercise their will and to weave out patterns of change not foreknown to Him. The patterns of change determined by the individuals have the required novelty, the Absolute wished to witness. The said patterns of change are thus creations in the real sense of the term.

The root and source of these creations are the individual will and its freedom. The individual will can create patterns of change in the inanimate and in the animate world because, as an individual it is a replica of the Absolute, though it has its limitations. The individual will can create because, being an entity which implies an autonomous power, its activities cannot be determined by the Absolute. An individual brings out new forms by changing the inanimate objects it comes in touch with. By exerting its influence on other living creatures, an individual also makes the animate world to undergo changes. The influence of individuals on each others' activities and on the material objects bring in changes in the world. And these changes create new forms and patterns unforeknown to the Absolute. These are also creations in the true sense of the term.

The creative activities of living beings may broadly be classified into three kinds on the basis of objects acted upon. A living being may act upon material objects, or upon other living beings, or he may act upon himself. The environment in which a living creature is placed is composed mainly of inanimate objects. Each movement of the said creature in connection with its collection of food, adjustment of body-temperature and pressure and protection of its life and body from malicious forces affects the environment and changes it. So, some changes the objects in the environment undergo are due to the exercise of individual will of the creatures. In course of evolution more and more developed species of creatures come into being who learn through their experiences the necessity of constructing houses to live in, constructing wells, ponds and canals for drinking water, constructing roads etc. for easy movement and producing utensils, weapons and such like things for obvious purposes. These are all matters related to the nutritional needs of living creatures. The species of creatures in the advanced stages of evolution thus affect the forms of inanimate Nature.

Coming to the activities of living creatures that affect each other we observe that all creatures consume either plants, or other creatures, or both, for their food. Here also in course of evolution developed species come into existences who feel the necessity of having a sure and regular supply of food for their sustenance. They therefore try to control the growth of living creatures, both plants and animals. They start cultivation for the production of food grains and vegetables. They create orchards to grow fruits. They control the growth and supply of fish through pisciculture. For a regular supply of meat they start poultry and control the growth of some birds and animals. They develop also a technique for cooking and preparation of food for better nutrition. They also develop techniques to make clothes from fibres etc. collected from some plants, and from skins and coats of animals, to protect their bodies from heat and cold.

Now, all activities of living beings affecting the inanimate objects and other living beings in their environment are based on their nutritional needs. These are activities of living beings required for the very sustenance of their lives. It being so, the basic pattern of these activities remains the same everywhere in the universe. Living beings have evolved on our planet earth in the solar system. Living beings might have evolved or may evolve in other parts of the universe as well, if the conditions are favourable for such evolution. The environmental conditions may be different in different areas of the universe and the body-structures and types of living beings may also not be similar everywhere but the basic requirements for sustenance of life being the same, the basic pattern of activities of living beings remain the same. These activities of living creatures affect the forms of inanimate Nature and the growth and development of living creatures themselves. These activities are also controlled by the Absolute but these are neither pre-determined by him nor foreknown to him. These activities are not pre-determined by the said creatures also. As such, these activities are creative in the real sense of the term.

As regards the activities of living creatures on themselves the most striking point we observe is that such activities are correlated to the stage of development of the creature, particularly to the development of its mind. When the mind of a creature is matured enough to discriminate between things properly, when it is matured enough to deliberate on situations properly and quickly and can guide the movements of the body in the correct line, it pays attention to itself. It studies its own ways of movement, analyses the mass of information it receives through the instrumentation of its own body and tries to study the forms and patterns of the universe. The mind studies the movements in the universe, goes on discovering the laws that guide those movements and utilizes the said laws, if possible, for the nutritional purposes of the body. In course of its analysis of materials received through the body, the mind comes to know the forms and patterns of objects in the universe. The mind comes in touch with the manifested universe through the sense organs of the body. The growth and development of sense-organs are correlated to the advancement of mind and the necessities of the body. The living creatures do not possess an equal number of sense organs. The least number of sense-organs in a living creature may be one, as in the case of least developed creatures like amoeba etc. The whole body of an amoeba functions as an organ of sense and an amoeba has only the sense of touch. In course of evolution, the living creatures develop more organs of sense and in case of human beings we find five such organs of sense—the eyes, the ears, the nose, the tongue and the skin. It cannot be said that five is the maximum number of sense-organs that a living creature may have. It is not impossible for living creatures to grow more organs to sense other aspects of the universe which are now unknown to us. Who knows, it might even have been done by creatures in some other areas of the universe. The main point for us to note here is that the sense-organs are grown and developed by a living creature if its mind feels the necessity of having that organ for its purposes of discrimination.

The mind comes in touch with the universe through the sense-organs and comes to know the various aspects of it. The mind comes to know the underlying forms and patterns in materials it receives through the sense-organs. In its advanced stages of development the mind not only comes to know but also feels the inner nature and rhythm of those forms and patterns. Experiences of such feelings thrill him because in such moments the mind comes in touch with an inner and deeper realm of the universe, superior and gigantic and sublime in comparison to what he ordinarily comes in touch with. The functioning of mind in such states of ecstasy becomes one with the workings of the manifested universe. The thrill, the ecstasy, the joy of such states generates also a feeling of awe and reverence in the mind towards the said rhythm and beauty. The mind therefore feels an urge to imitate that rhythm and beauty in different media. These imitations are the artistic creations of living creatures like men.

The mind comes to know the colour, the shape and the movements of objects in the universe. It enjoys and appreciates them and directs the body to record them by imitation on the media available for future enjoyments. The living creature therefore paints pictures on different media and records the colour, the shape and the movements of objects. To record beauty and harmony in the shapes of objects they carve out statues and they imitate the movements in Nature in dances. Due to movements of inanimate objects and living creatures, sound, rhythm and melodies are generated in Nature. The mind enjoys the harmony and beauty of sound and directs the body to create similar sounds and the activities of creatures towards that end result in music. The mind senses the fragrance in various objects of Nature and directs the body to create articles for such fragrances. A living creature also develops the art and technique of cooking delicious food to please the tongue. All the above activities of a living creature are creative and artistic in nature. All these artistic activities of a creature are based on his awareness of the quality, the form and the rhythm in Nature. The artistic sense in a creature modifies and beautifies all objects created by him. The sense of beauty impels the creature to beautify even the articles of his food. Various other articles are created which serve no nutritional purposes but only beautify the surroundings. The mind apprehends the movement and life of Nature and the creatures that live in it. This apprehension connects him with the general stream of life in his surroundings. The movement and rhythm of this stream of life makes him aware of the workings of a vast and animate world. He creates the rhythm and beauty underlying such workings and imitates it by creating a world of literature.

Though artistic creations are originated in the urges of living creatures to imitate the forms, the patterns, the rhythm, the harmonies and the beauties underlying the movement of inanimate and animate objects in the universe, they do not remain confined in such activities only. When a living creature comes in touch with a beautiful object he does not necessarily feel impressed. He may not appreciate the form and pattern in it. For example, a lower animal, an infant or a man of rustic nature does not feel impressed with our so-called beautiful objects. Appreciation of beauty therefore depends not only on the beautiful objects but also on the perceiver. Appreciation of beauty means a response on the part of the perceiver towards the form etc. of an object or objects. An urge to respond to the beauty of an object is not generated in the feelings of wonder, for, in that case a lower animal, an infant or a man of rustic nature also would have responded to the so-called beauty of an object as they experience the feelings of wonder quite frequently. Appreciation of beauty is a mental phenomenon and it depends, according to us, on the mental development of a creature. Does it depend on the knowledge acquired by a creature? No, for in that case only the learned beings would have the sense of beauty. It is a fact that there are mathematicians and scientists, historians and economists, who have very poor or even no sense of beauty. Does then the sense of beauty depend on a comprehensive knowledge about the universe? No, because there are philosophers who have no keen sense of beauty and artists who are illiterate. Knowledge, in the general sense of the term, is not therefore a pre-requisite for the development of a sense of beauty. Where then does the sense of beauty originate? According to us, the sense of beauty is nothing but a feeling of unison generated in the mind of a creature between the inner workings of an object of beauty and the inner workings of his mind. Apprehension of beauty is a bi-polar process. It depends no doubt on the beauty of the object, because any and every object cannot generate this sense in the mind of a creature. It depends also on the state or condition of the mind of a creature that is to respond, because any and every creature cannot do it. The form, the pattern and the rhythm of an object acts on the mind of a creature, and the mind of the creature reacts in the form of a response. Now, why only a particular type of mind responds to an object of beauty? It is because only such minds have in them the principle required for effecting a response to beauty. What can that principle be? Whatever may that principle be it must at least be similar to or identical with some principle that exists in the object of a beauty, for, otherwise there cannot be correspondence between the object of beauty and the mind that responds to it. If there is nothing common to them they cannot, according to us, act or react on each other. According to us, the element common to both an object of beauty and the mind that responds to it, is the principle governing the inner workings of them both, for, what else is common to them? To be aware of the beauty of an object and to appreciate it is to be aware of the form, the pattern and the rhythm brought there out by its inner workings or movements. Now, the form, pattern and rhythm are not properties common to an object of beauty and the mind of a creature, because the mind itself has no such properties. The only property common to them can therefore be the principle underlying the inner movements in them. To be aware of beauty is thus to be aware of the inner workings of one's own mind first and then, through the mind, the form, the pattern and the rhythm actually observed in the object of beauty. The form etc. observed in the object of beauty, are, in a sense, present in the mind of the observer also because the inner movements in the object of beauty and in the mind being similar, the said forms etc. could be felt by the observer. Awareness of the beauty of an object is thus a feeling of unison between the inner movements of Nature and the inner movements of the mind. An object of beauty makes a responsive mind aware of its own inner movements and also of the inner movements of Nature. This awareness thus connects an observer with entities other than himself or with a realm greater than that of his own. He is therefore thrilled with an ecstatic joy. He appreciates the beauty of the object and feels inspired to imitate such object through artistic creations.

Artistic creations are based on life and Nature. What an artist creates is a perfect or imperfect imitation of the form, the pattern and the rhythm of a beautiful object. The object was reflected in the consciousness of the artist or in his mind and was depicted thereafter on a medium. The object of art is based on the said reflection. It is the translated form of the object of beauty in the mind of the artist. Whatever objects an artist creates, it is an expression of his feelings which is a part of his life. An artistic creation is thus rooted in life. An object of art is also based on Nature, in the sense that the form, the pattern and the rhythm depicted therein are based on a beautiful object of Nature.

Creations of art do not of course remain confined in this basic form for ever. There is evolution in this sphere also. The form of art depends on the nature and standard of the artists' apprehension of inner movements in a beautiful object of Nature. In its crudest form, a piece of art is nothing but a more or less perfect photograph, or replica, or imitation of a beautiful object of Nature. At the lowest stage of his mental development an artist tries only to imitate the beauty of an object in as perfect a way as he can. His apprehension of the inner movements in an object of beauty and the similar movements in his own mind are only vague and superficial, not deep and perfect. He has only managed to catch the rhythm of movements somehow. His way of expressing the said rhythm is also not perfect and he has no mastery over the technique he follows. He looks at an object of beauty, is stirred by it, feels an ecstatic joy and tries to imitate it by drawing its picture in his crude and staccato manner. The picture is only an imperfect replica of the object of beauty and an imperfect expression of what he apprehends of it. The artist hears a rap of appealing sound, tries to imitate it and creates a melodious tune which is only on imperfect copy of what he has apprehended of the former. In course of his mental development the artist goes on developing his artistic faculties by penetrating deeper into the inner movements in an object of beauty and into the inner movements of his mind and also by perfecting his technique of expression.

When an artist is able to go deeper into the inner movements of an object of beauty through similar movements of his own mind he is at a higher stage of the apprehension of beauty. When the technique of his expression is also developed he becomes an artist of high order. But an artist need not always keep his activities confined in imitations only. When he has apprehended quite a number of beautiful objects and has apprehended the very principles of movements which create those beautiful forms, patterns and rhythm the artist need not imitate the natural objects of beauty. As the very principles of movements in the beautiful objects are known to him he can now create his own objects of beauty. He may even create a form, a pattern and a rhythm of his own which will exhibit the inner movements of Nature all the same and the object created may have a life of its own. The object created may appear to be something not found anywhere in Nature but as it is rooted in the very principles of movements which produce the forms, the patterns and the movements in natural objects, it is not devoid of life in it. Though it is created by a creature like man, though it is not like any natural object of beauty, it will have life like other beautiful objects of Nature. The artist who can create in this manner is a master artist of super order. He has penetrated deep into the realm of beauty. He has apprehended the inner movements that make the natural objects of beauty through the apprehension of similar or identical movements in his own mind and has also developed a technique to control such movements in order to create objects of beauty of his own. As the movements are within his control he can orient them in his own way and can create new forms, new patterns and new rhythm. He has attained such a stage of mental development that he can control and mould the inner movements of his mind in all possible ways to create objects of beauty which have life-like forms, patterns and rhythm. As a creator he is almost like the Absolute who also moulds and remoulds the inner movements of objects to create forms, patterns and rhythm and creates objects of beauty. The artist who has that keen sense of apprehending the inner movements behind all forms, patterns and rhythm in Nature and in his own mind, the artist who apprehends even the possible courses of development in Nature almost becomes one and identical with Nature or the Absolute. The feeling and experience of the artist in such a state is religious, because in ecstatic moments of religious experience also one feels himself one and identical with the Absolute or God.

Let us now examine the ontological status of artistic creations. Are they only products of imagination with no reality behind them? When a painter paints a picture does he only permute and combine various colours and shapes of objects? Does the picture represent only the imagination of a painter's mind that would never have been brought out by Nature because it was beyond the region of possibilities? When a novelist draws pictures of a society and creates characters in it, does he depict something impossible for Nature to have brought out into existence? When a musician produces a piece of music does he only combine some notes of sound and does the piece only bear stamp of the musician's imagination having no reference to reality? Does the piece of music represent nothing that could be brought out by Nature?

We have shown earlier in this chapter that artistic creations are based on the principles that guide the inner movements of objects in Nature. An artist apprehends the inner movements in Nature which happen to be the same in his own mind because he himself is nothing but an object of the same Nature. Following the principles that guide those movements, the artist moulds and remoulds the inner movements in his own mind and creates new forms, new patterns and new rhythm as is done by Nature herself. The artist, of course, is to have a mastery over his knowledge of principles underlying those movements. He must have acquired knowledge enough to have a command over them. He is also to develop and master a technique of applying those principles to create new forms, new patterns and new rhythm. Mastery over the principles underlying inner movements in Nature and over those in his own mind makes the artist apprehend the alternative and possible courses of movements that could be adopted by Nature herself. These courses must have coherence in them. The artist can therefore adopt one of such courses and can weave out new forms, new patterns and new rhythm accordingly. His creation will therefore have coherence in it and will represent something natural. An artistic creation of this kind will seem to be something brought out by Nature herself. It will have a real form, a real pattern and a real rhythm. If, on the other hand, a person has not apprehended the inner movements of objects in Nature, if he has not felt properly the inner movements of his mind, he can at best have the knowledge of some discrete elements of the forms and patterns of objects in Nature. Even if such a person develops a technique to express his feelings in a medium, his creation cannot have any coherence and rhythm in it. Such a creation can at best be a poor replica of objects in Nature. It cannot have a coherent form, pattern and rhythm in it and as such, it can have no beauty. A creation of this kind cannot be a source of ecstatic joy because it has no life or growth.

Now, a creation is real when it conforms to a form, pattern and rhythm in Nature, either actual or possible. Whatever object is brought out by Nature, it is real. Whatever object could be brought out by Nature also must be real because it would have been possible for her to bring it out. It would have been possible for Nature to bring out the said object because the object existed in Nature in an implicit form. In the case of a creature also, whatever is created by it is real because it has somehow been possible for the creature to bring the object out into existence. Thus, whatever is brought out by Nature or is created by a living being are real because they have somehow come into being. There cannot but be coherence in objects brought out by Nature because, being a manifestation of the Absolute, Nature covers all and all objects are somehow consumed by and adjusted in it. There can however be objects brought out by creatures which have no coherent relations amongst them, though with reference to the Absolute they are also real because they are also covered and consumed by the Absolute. From the standpoint of the Absolute, such creations of living beings are real all the same but they have no beauty because of the absence of a coherent form, pattern and rhythm in them.

As all objects brought out by the Absolute have a coherent form, pattern and rhythm in them they are beautiful. An artist's model can therefore be anything and everything brought out by Nature. But then why only some objects of Nature look beautiful to a creature? It is because a creature, who has limitations in his organs and limitations in his method of apprehension, cannot have the comprehensive outlook necessary to apprehend the form, the pattern and the rhythm in all objects of Nature. To a creature, only those objects of Nature appear as beautiful which have their forms, patterns and rhythm coherent and complete in themselves and which need no broader perspective to exhibit coherence in their forms, patterns and rhythm. A flower looks beautiful because its form, pattern and rhythm are coherent and complete in themselves. A landscape is beautiful because its form, pattern and rhythm are coherent and complete in themselves. But any area of earth, any part of a broken hill, any limb of a creature, will not look beautiful because their forms, patterns etc. are not coherent and complete in themselves. Nothing is ugly in this universe because, looked at from the viewpoint of the Absolute, all objects are coherent and rhythmic. An object looks ugly if the coherence underlying it is not complete in itself, if it requires a broader perspective to complete its coherent form, and if a creature does not look at the object from that perspective.

When an artist aims at the creation of a form, pattern and rhythm underlying an actual or possible tree, the tree he creates must truly represent that very form, pattern and rhythm. If the tree created fails to represent the conceived form, pattern and rhythm, then this piece of art will be a failure and it will not be a source of ecstatic joy for which it was created. A piece of art fails to be a source of ecstatic joy if either the artist fails to apprehend and conceive properly the real form, pattern and rhythm underlying an actual or possible object, or, if the artist did conceive them properly but failed to express and depict that form, pattern and rhythm. Thus artistic creations either conform to the actual or possible courses of development in Nature and in the universe, in which case these are successful creations of art, or, these creations may not conform to any courses of development in Nature and may only be cases of failure. However, in both the said cases the objects created in art are real from the standpoint of the Absolute, because in both the cases the objects created are adjusted in Nature and consumed by the Absolute. In the former case, the objects created by an artist are beautiful both to the creature and to the Absolute and in the latter case the objects created are real and as such beautiful only to the Absolute.

The Absolute manifests a bulk of its energy to bring out entities that may produce something unforeknown to Him. The evolution of living creatures brings these entities out into being. These creatures enjoy freedom in the exercise of their will and create objects through their economic, artistic, philosophic and religious activities and these objects or entities are unforeknown to the Absolute. The economic activities are the activities of living beings which are performed by them for the maintenance of their bodies and for easing the hardships they face in life. The artistic activities are based on their apprehension of the form, pattern and rhythm in the objects of the universe. These are attempts on the part of living creatures to imitate Nature and to create objects and entities charging them with the forms, patterns and rhythm, either actual or possible, that may serve as sources of ecstatic joy for them. The philosophical activities are attempts to know the principles that work behind the evolution of the universe and to have a comprehensive conception of the universe as a whole. Philosophical activities do not result in the creation of concrete entities like those that result in other activities. Philosophical activities which are apprehensive in nature result in the creation of mental images, real or unreal, and ultimately terminate in the religious activities which are attempts of creatures to overcome the barriers of their limitations and to make themselves one with the Highest or the Absolute. These are attempts for realisation of oneself in the correct perspective and are carried out through the activities, economic and artistic. The said activities, involved in religion, are of an order different from the order of economic activities carried out for the purpose of maintenance of the body and mind and different from the order of artistic activities carried out for the enjoyment of beauty. The religious activities are oriented towards an union with the Highest and such activities bear a sense of the Holy.

All creations of living beings are unforeknown to the Absolute and as such they display novelty before him. The artistic, the philosophic and the religious activities of creatures help them to come in touch with the inner movements of Nature and the universe. This is the natural course of development for a living being. We have shown earlier that the bulk of energy manifested by the Absolute moves on undergoing development and attains the state of inorganic and organic compounds. We also say that this is the natural and normal course of development for a bulk of manifested energy. The manifestation of consciousness in organic compounds has also been shown by us as a natural course of development. The organic compounds fulfil no purposes of the Absolute unless they are individuated properly. Consciousness, therefore, appears in an organic compound to individuate it. We have also shown that individuation is nothing but a principle of discrimination and integration of knowledge or rather the differentiation and integration of consciousness. To differentiate consciousness and to integrate the differentiated consciousness is the natural course of development for an individuated entity or a living being. A living being comes in touch with the universe through its sense organs and acquires knowledge of it by analysing the sense data in its mind. This analysis is a process of differentiation, integration and ultimate extension of knowledge. Extension of knowledge means an expansion of one's consciousness. In course of expansion, the consciousness of a creature ultimately brings the whole universe within its fold. A living being acquires at that state infinite knowledge, and being now an entity having infinite consciousness it becomes one with the Absolute. Thus the natural course of development for a living being is to be one with the Absolute. All activities of a living creature that are directed towards his union with the Absolute are therefore natural and normal in the scheme of the universe.

The economic activities of a living creature are meant for the purpose of sustenance of his body and mind and these activities conform to the will of the Absolute. When such activities do not sustain the body and mind of a living being, these are to be considered as forces that act against His will. The artistic activities of living creatures are meant for apprehension of rhythm and beauty in the inner movements of Nature and the universe. Such activities that do not lead to fulfilment of the intended purposes therefore get frustrated. The philosophical activities are meant for the purpose of a rational understanding of the principles and entities that work behind the evolution of the universe and such activities conform to the will of the Absolute. Religious activities of living beings are supposed to effect an union of living beings with the Absolute. Such activities that do not lead to an union of living beings with the Absolute are therefore forces that act against the natural course of events. All activities of living beings that conform to the will of the Absolute get accelerated because the natural course of events also works behind them as an additional force. He who helps Nature is helped by Her. A creature in the natural course of his developments is therefore sure to reach its goal.

We have shown that the Absolute is not a creator in the true sense of the term. Being the source of all that are manifested in the universe, the Absolute cannot create anything. Nothing, from his standpoint, is new and unforeknown and as such he cannot be a creator. The creatures, on the other hand, can be and are actually creators. They are creators because they can exercise their will freely and bring entities out into being the state of which are unforeknown to themselves and to the Absolute. The said entities are new to the creatures because they could not have a foreknowledge of them. These entities are new to the Absolute also, in the sense that though the Absolute had a foreknowledge of the manifestation of many possible entities he could not determine which particular entity the creatures would bring out into being, because, such determination depended on the creatures who had the freedom to exercise their will. A creature's freedom of the will, however limited it may be, gives him a private sphere of his own of which even the Absolute can have no foreknowledge. It is so, because this privilege of creatures to have private spheres of their own has been granted by the Absolute Himself.

Though not a creator, the Absolute is an artist of the highest possible order. We have shown that art is a reproduction of forms, patterns and rhythm in Nature, either actual or possible. The Absolute, in this sense, cannot be an artist because he cannot reproduce the forms, patterns and rhythm of his own constitution. He cannot also have a medium to work upon because there is nothing outside him. But the Absolute is an artist in the sense that He is beauty itself. If beauty means a coherent form, pattern and rhythm, then the Absolute is all-beauty. If to be an artist is to apprehend the forms, patterns and rhythm underlying the objects of the universe, then the Absolute is an artist, par excellence. An artist, from the level of creatures, apprehends beauty in some objects of Nature only. He cannot apprehend beauty in all objects and some objects appear to him as even ugly. But the so-called ugly objects and disharmonies of Nature appear to be so when looked at from the standpoint of creatures only. The various limitations in a creature act as barriers to his having a comprehensive view of the universe as a whole and as such the coherent structure of the universe, where all objects have their beauty and harmony, remain veiled to him. But being an entity having no limitations, the Absolute is the said harmony itself, because all objects and their forms, patterns and rhythm are adjusted in him harmoniously. The Absolute is thus a super-artist in this sense. All ugliness, all discord and all disharmonies are adjusted, absorbed and harmonized in him. He is therefore all-beauty and all-art.

All that happens in the mind of a creature are items or elements of his consciousness. There are four states or phases in the progress of a creature's mind corresponding to the four different types of his activities. In its first phase, corresponding to the economic activities of a creature, the mind is discriminative. Mainly for the purposes of the body, i.e. for its maintenance and growth, the mind discriminates between things it comes in touch with through the sense organs. This discrimination and deliberation is necessary for the mind because it is to guide the body in its movements for sustenance and growth. The mind receives information about a creature's environment, analyses the same, deliberates on the matter and guides the body to move accordingly. When the movements of the body bring the desired results, the mind is satisfied. If otherwise, the mind is disappointed. It then analyses the situation again, discriminates further and guides the body again to have the desired result. The process continues till the purposes of the body are served and the mind feels satisfied. Discrimination in this phase of the mind continues so long as consciousness and its centre, the mind, remain attached to a body. The second phase of the mind is a phase of integration and it corresponds to the artistic activities of a creature. In this phase the mind does neither analyse nor discriminate. The mind rather synthesizes in this phase what it analysed in the first phase. Instead of being engaged in discrimination and deliberation, the mind here apprehends and feels what it had previously deliberated upon. If in the first phase the mind collects information about Nature, then in the second phase it absorbs that information. For mind, to reach at the second phase from its first phase is to make progress and it implies an enlargement or expansion of individual consciousness. In the third phase, the mind of a creature analyses its knowledge, acquired in the first two phases, in order to find out the universal principles underlying it. The mind in this phase tries also to determine the first principle or principles that worked behind the evolution of the universe. This phase of the mind corresponds to the philosophical activities of a creature which are discriminative and analytical in nature. Here the mind develops a comprehensive conception of the universe as a whole. Consciousness of a creature in this phase is enlarged further. In the last phase, corresponding to the religious activities of a creature, the mind synthesises or integrates its philosophical knowledge, acquired in the third phase. In this phase the mind is no longer engaged in discrimination or deliberation. The mind here apprehends, absorbs and assimilates its philosophical knowledge and convictions. The consciousness of a creature at this phase covers the whole universe and so it becomes infinite. It is now one with the Absolute.

Chapter 7
Struggles in Life and a Note on the Moral Life of Men

We have stated earlier that the first principle behind our universe is an infinite consciousness which also implies infinite energy in it. We have named this entity, the Absolute. We have shown that the very existence of infinite consciousness which implies change in it requires energy for it. Energy and change are thus the contents of this prime and infinite consciousness. Infinite consciousness cannot exist unless it undergoes change and it cannot undergo change unless there is energy for the purpose. Again, consciousness means a principle which is conscious of something. As everything may be reflected in consciousness, so consciousness itself will also be reflected in it. Consciousness itself also is thus a content of infinite consciousness.

Prime-consciousness is infinite because its contents are not limited in extent. Energy, in prime-consciousness, is unlimited because the said consciousness, as an existent entity, cannot cease to exist and as such, will continue to exist for ever spending energy. Energy cannot therefore be limited in quantity, for if it is so, prime-consciousness will have to cease to exist at a certain stage. The existence of prime-consciousness for an infinite period implies an infinite change it undergoes. As there is nothing beyond prime-consciousness, it cannot have a boundary. Prime-consciousness is therefore infinite. This infinite consciousness, which is conscious of infinite energy, infinite change and of itself, is the first principle behind our universe and we have named it the Absolute.

There is nothing beyond the Absolute because, according to us, consciousness, energy and change are the stuff or matrix to which all entities of the universe are reducible. Let us now see whether or not these three principles are reducible to each other. Now, energy, we know, is the capacity for doing work. Energy does not necessarily mean work actually done. It may also mean the potentiality of doing work. In the case of work actually done change is involved. But in the case of potential work, change is not involved. Thus change is a principle implied by energy but separable from it. Change is therefore a principle separate from the principle of energy and as such they are not reducible to each other. As for consciousness it may be stated that it exists and as there are no reasons for it to cease to exist, it must undergo change and expend energy for the purpose. It may therefore be said that consciousness has in it energy and the principle of change, though in an implicit form. But being the contents of consciousness, energy and change cannot be reduced to or equated with it. Moreover, consciousness is a principle of discrimination whereas energy and change are not so. Consciousness, energy and change are thus not reducible to each other. Energy and the principle of change exist in consciousness, though in the implicit form and these are the contents of consciousness. This entity is the Absolute of our conception.

Is the Absolute an entity beyond time and space? Obviously, it is not. Continuity of the Absolute's existence implies the presence of change or time in it. Even if we think that before continuing to exist the Absolute did not undergo any change and was beyond time, the condition of its having the principle of change in it in an implicit form makes it an entity not beyond time. The position of course is that the Absolute is not in time, rather that time is in the Absolute. Similar is the case with space. We have shown earlier that manifestation of energy means a separation of it from the original mass of integrated energy that lay implicit in the Absolute. This process of manifestation makes the said bulk of energy come to a separate level of existence. The principle of differentiation works also on the mass of manifested energy and goes on making its elements separated from each other. Separation of the said elements from each other makes it necessary for each of such elements to have an exclusive area of its own. These exclusive areas are what we mean by space. Space thus comes out of the Absolute when a bulk of energy is manifested, and space, as a principle, is thus derived from the principle of manifestation. As the Absolute can manifest a bulk of its energy, if it so wills, it may be said that space lies in the Absolute in an implicit form. The Absolute is not therefore an entity beyond space. As in the case of time so also in the case of space it may of course be said that the Absolute is not in space but rather that the space is in the Absolute. It may here be asked: If manifestation of energy is the root of our conception of space, then does space not come out when, in order to continue to exist, the Absolute brings out energy from its implicit to its explicit form and spends it for the purpose? Our answer to this question is: the energy in question is being manifested to effect some change in consciousness and consciousness requires no space to be in. When the Absolute continues to exist by spending energy for the purpose it does not bring energy from one level of existence to another. The changes in energy due to its movement from the implicit to the explicit state are all effected in consciousness only and as such energy in course of this process does not get separated from the level of consciousness. Space need not therefore be brought out for the purpose. A later stage of the Absolute is nothing but its previous stage plus the energy it has spent to maintain the continuity of its existence. If it is not so, then there cannot be any difference between the Absolute of two different stages. Existence of energy in the Absolute has made change in the latter possible. Energy spent is thus change which the Absolute has absorbed in it. But does this energy come out of the integrated mass where it existed so long? No, because this energy only makes the continuity of existence possible for the Absolute and does neither get separated from it nor move on to a separate level of existence. Moreover, for the maintenance of continuity in its existence the Absolute does not manifest energy, in the strict sense of the term. Here the Absolute only spends energy to maintain continuity in its existence. It does not here bring out its energy in a bulk. The difference between cases of the Absolute in continuing to exist and its bringing out a bulk of energy from within itself is that in the former case energy is spent and in the latter it is manifested. In the former case, energy is absorbed in the Absolute in the form of change it undergoes, and in the latter energy is transferred to a different level of existence. There lies surely a difference between an expenditure of energy and a manifestation of energy. In the case of the Absolute's continuity of existence energy is converted to a change in consciousness. Energy here is not brought out as a specific entity. Energy exists in the Absolute implicitly and a part of it makes the Absolute to continue to exist. The implicit energy and its part in question do therefore lie in the same level of existence. But when a bulk of energy is manifested by the Absolute, the said energy is brought out as a specific entity and as such it goes to a different level of existence. In the Absolute, which according to us, is energy-consciousness-change, the constituent elements are not separated from each other. The elements here lie integrated and each one implies the other two. But in the case of manifestation of energy, the bulk of energy is separated from this integrated mass and it assumes a specific form of its own, the form of being in space.

There is nothing beyond the Absolute and as such nothing can go beyond it. Energy, spent or not, is within the fold of the Absolute. Energy, manifested or not, also does exist within the Absolute. As change is the essence of existence, there is change not only in consciousness but also in energy. Energy, whatever level of existence it is in, does undergo change. Energy, implicit or explicit, does never remain in the same state. What then is the relation between energy and change? Change, we know, is present in the very constitution of energy and consciousness. Being a principle that requires a medium to work upon, change cannot be identical with its media, energy and consciousness. Change is a principle which is other than the principles of energy and of consciousness though the latter principles imply the former. We can derive change from the principles of energy and consciousness but cannot derive energy or consciousness from change because change does not necessarily mean that it is to work on energy and consciousness.

The three principles of energy, consciousness and change in one entity is our concept of the Absolute. Can there be a creator of the Absolute? We give a negative answer to this question. The Absolute being infinite there cannot be anything beyond it. Therefore, if we conceive of a creator of the Absolute, that creator must either be beyond the Absolute, or, identical with the Absolute, or, within the Absolute. The first alternative is absurd here because being infinite the Absolute has no boundary. The third alternative is also absurd because the creator in that case will be an entity lesser than its creation. The second alternative must therefore be the fact and what it means is that there can be no entity other than the Absolute. The Absolute was, is and will be.

The Absolute can exist for ever without manifesting a bulk of His implicit energy. But, for reasons we have shown earlier, the Absolute wills to manifest a bulk of His energy and He does so. The will of the Absolute is thus the moving power behind the workings of the manifested universe. If in the manifested world the individuals can have their will, there cannot be any contradiction in believing that the Absolute who is the source of this world, has a will and that He exercises that will.

We have shown earlier that manifestation is a principle which means differentiation of an entity from the mass of integrated elements in which the entity existed undifferentiated. The principle of manifestation will continue to operate on the mass of energy so long as the Absolute does not will to draw the said energy back to the integrated mass, or rather to itself. The principle of differentiation reinforced by the principle of change that operates on every existent entity and the principles specific for energy, or the principles relevant to the very existence of energy, generate a force. This force causes the universe to move and to evolve. It is the élan vital, in the language of Bergson. It starts in the manifestation of energy and ends in its absorption in the Absolute.

The bulk of energy manifested by the Absolute assumes a distinguishable and specific form of its own, for, that is the meaning of manifestation. The principle of manifestation now works on this specified energy causing its various elements to assume specific forms of their own. The bulk of energy is thus ultimately converted to the indivisible units or corpuscles of energy. Movements in the corpuscles of energy makes them to come in contact with each other and the principles of differentiation and consequent integration make them to form atoms, molecules, elements and compounds of elements. Then the principles of differentiation and integration assume the form of individuation. This principle of individuation works on some chemical compounds resulting in the appearance of life, or, consciousness, and mind in them. Consciousness works through the organic compound, or, the body in which it does appear and goes on acquiring knowledge of the universe till it becomes infinite consciousness, or one and identical with the Absolute.

Due to manifestation, energy comes to a different level of existence which, according to us, is space. Space, according to us, is that principle which makes possible the exclusion of entities from each other. The bulk of quantity of energy manifested by the Absolute therefore determines the extent of space necessary for its accommodation. More space is necessary for more energy manifested and less for the manifestation of less energy. As it has been possible for the Absolute to manifest a bulk of its implicit energy so there is no contradiction if we say that the Absolute can withdraw a part of such energy or can manifest more energy. The extent of space will then be decreased or increased accordingly.

The elements of energy, separated from each other, now bring in a relation amongst themselves, the relation of proximity to and distance from each other. When the bulk of energy was in an unmanifested state, it was an integral element of the Absolute and as such the question of proximity or non-proximity did not arise at all. After manifestation the said energy came to a different level of existence. This level of existence to contain the manifested energy is space, according to us. Being only a level of existence to accommodate some elements of the Absolute, space is within the fold of the Absolute. All possible levels of existence do surely exist in the Absolute and space is one of them. Being only a possible state of the existence of energy, space is not a physical entity. It cannot therefore have a boundary. Though without a boundary, space will have an extent depending on the quantity or bulk of energy manifested. There cannot be empty space which has no connection with the corpuscles of energy. Space, being only a principle of exclusion for the corpuscles from each other, cannot be free from such corpuscles, for, in that case space will be a principle of exclusion with nothing to exclude.

Manifestation of energy and its consequent differentiation cause energy to form corpuscles separated from each other. As each corpuscle has a boundary of its own it excludes itself from other such corpuscles and tries to maintain the integrity of its structure. Thus differentiation of energy also means its integration. The differentiation of energy into corpuscles and the integration of these corpuscles generate forces, both inwards and outwards. The force inwards is the pull towards the centre of a corpuscle of energy. It is so because the principle of differentiation causes a corpuscle of energy to separate itself from all other entities and to keep its boundary intact. Now, a pull inwards, i.e. towards the centre of a corpuscle means also a pull outwards, i.e. away from the centre of an adjacent corpuscle. To keep its boundary intact a corpuscle is to resist all forces outside its boundary that may threaten to break it up. Thus there are forces and counter-forces working on the corpuscles of energy and these forces are reinforced by the principle of change. The said attraction and repulsion of the corpuscles of energy create a commotion amongst them and new entities begin to appear. This commotion is more or less according as more or less corpuscles lie near to each other.

The commotion in manifested energy creates movement in the corpuscles of energy. This movement of the corpuscles of energy is possible if only the medium, in which they are, allows such movement. Being the medium in which the corpuscles of energy are, space therefore must allow the said corpuscles to move. And what it all means is that space in such cases gets extended. This extension of space will be impossible if space happens to be a physical entity. If space is a physical entity then it must have a boundary. But then, beyond the boundary of this space there must again be space to make the movement of corpuscles possible. Space cannot therefore have a boundary and as such it cannot be a physical entity. Can then the space be a mental entity? Not being physical, space cannot but be mental. Space is nothing but extension coexistent with the extension of energy. Space has no independent existence of its own. Its existence depends on the existence of energy manifested. The extent necessary for the accommodation of all energy is the extent of space. If due to forces and counter-forces, the corpuscles of energy move away from each other and more area is necessary to accommodate them, then space gets extended. If, on the other hand, the corpuscles of energy lessen their distances from each other, then space needs a lesser extension. Space, in that case, will have a lesser extension, according to us. Just after manifestation, energy was yet to be differentiated into corpuscles and as such a lesser area was necessary for its accommodation. Space had at that stage a lesser extension than what it has now. Differentiation of energy into corpuscles and the consequent workings of forces and counter-forces on those corpuscles made them to move away from each other and space extended itself accordingly to accommodate them. The extension of space is identical with the extension necessary for the accommodation of all the corpuscles of energy in their various positions. Thus space is not a separate physical entity. It is only a level of existence conceived by the mind to accommodate manifested energy. This point is further proved by the fact that space is not necessary to accommodate consciousness and mind. Consciousness and mind, being entities in a level of existence different from that of manifested energy, cannot be accommodated in space. Consciousness and mind which are able to conceive different levels of existence conceive space as a level of existence for manifested energy.

As regards time, we have said earlier that it is nothing other than change. Change is a principle implied in the very existence of consciousness and energy and these three principles taken together is the Absolute in our hypothesis. Now, time is generally believed to be an entity other than change. It is believed that all entities in the universe undergo change and it is possible because of the existence of time as a separate entity. But actually time, so conceived, is nothing other than the process implied in change. We make here a distinction between change and the process of change. By the word 'change' we mean the altered position of an entity. 'Change' may be slight, when the state of an entity is only slightly altered, or it may be radical, when the state of the entity is altered wholly. But by the words 'process of change' we do not mean the altered position of entities but the speed at which such alteration is effected. Change and the process of change thus do not mean the same thing. There may be a slight change effected very slowly and a radical change effected quickly. There may also be a slight change effected quickly and a radical change effected very slowly. But though there are differences in their meanings, change and the process of change imply each other. Change implies a speed at which it is effected and vice-versa. The one cannot be separated from the other. Now, the general concept of time is nothing but the process of change or the speed at which the change is effected. As there are changes in all existent entities of the universe it is believed that the said entities exist in time, as if time is an entity independent of them. But actually time as a separate entity other than the change is mere abstraction made in order to indicate and in some sense to perceive the speed of changes effected on entities. Time has not only been separated from change but has also been standardised by making divisions in it in the form of seconds, minutes etc. But this division and standardisation cannot but be arbitrary and relative because nothing being stationary in the universe the speed of change in an entity can neither be determined nor standardised. As there cannot be a speed of change unless there is a change, so also there cannot be time without change effected on an entity. Though it is believed that changes effected in entities are in time, the fact, as per our analysis, is that there is time in the said changes. That time is nothing but the speed of change undergone by an entity is also proved by our apprehension of it. When we perceive a radical change in some object previously known to us, we feel that a long period of time has elapsed since our previous perception of it. If instead the change in the object is slight we feel that a short period of time has elapsed. This shows that our idea of time is nothing but the speed at which changes are effected on an object. In the case of a radical change we are amazed at the work of time because the actual changes undergone by the object is much more than what it should have been according to our pre-conceived idea regarding the speed of changes that the object might normally have undergone. In the case of a slight change in an object, it is otherwise. In such a case we compare the actual change in the object with our pre-conceived standard of changes and believe that as the changes are only slight, a lesser period of time has probably elapsed. In the case of an enjoyable event we get ourselves absorbed in it and forget all about time. Then suddenly we become conscious that the event will soon come to its end. This makes us feel that changes in the event are rapid, or that the period of time elapsed is longer than the actual. In the case of an experience that bore us it is just the reverse. In such a case we wish to but cannot cross over the monotonous experience soon. We therefore feel that changes in the event are too slow, or that the period of time elapsed is shorter than the actual. All these phenomena concerning our feelings of time show that we conceive of time as the speed at which the entities undergo changes. Our thoughts and feelings about time are thus not the same. While thinking we consider time as an entity other than and separated from the changes of entities whereas in our feelings we find time to be the speed at which entities undergo changes. As per our hypothesis, our feelings about time are nearer to truth than those of our thoughts about it.

We have shown earlier that manifested energy and consciousness are at different levels of existence and that while manifested energy is spatial, consciousness is not so. But as existent entities there are changes in both consciousness and energy. As change implies also a process of change or rather a speed at which entities undergo change, time is involved in the existence of everything in the universe. But how is time related to space? There can be time without space as it is in the case of consciousness. Consciousness is not spatial but it has time in it. Space, on the other hand, cannot but be in time, because, being a level of existence, space contains manifested energy which exists and as such is temporal. Space is coexistent in extent with energy manifested. Then due to differentiation and the consequent phenomena of attraction and repulsion, the elements of manifested energy move away from each other and a relation of distance ensues amongst them. The total extent covering the corpuscles of energy, according to us, is the extent of space. As the corpuscles of energy continue to act and react on each other, their distances from each other do not remain the same and as such, the extent of space also does not remain the same. Thus changes in the corpuscles of energy make space also to undergo change. This further means that space has time in it. Thus there may be time without space but not space without time. In fact, space and time meet each other in manifested energy. Space is an outcome of the manifestation of energy. Manifestation of energy brings out space as a separate level of existence for the said energy. Could energy be manifested in a way other than by putting it in a different level of existence, called space? We do not think so, because, to manifest energy means to differentiate it from the integrated mass of implicit elements and for this purpose energy must be shifted to a different level of existence, whatever the nature of that level may be.

We have shown elaborately how the principle of manifestation works on energy and how there appears in space corpuscles of energy which act on each other and balancing the forces involved form their combinations like electrons, protons etc. Balancing similarly the forces and counter-forces involved in the movements of electrons, protons etc. atoms, molecules, elements and the compounds of elements come into existence. The corpuscles of energy are stable in form because, being indivisible units of energy, they undergo no further division and so keep their integrated form intact. But the compounds of such corpuscles are divisible and as such may not be stable in form. The balance of forces arrived at in them may be lost due to other external forces acting on them. The organization and balance of forces go on reaching higher and higher stages in more and more complex combinations of the corpuscles of energy but such organizations may be lost due to external forces acting on them. Further, differentiation of such an organization thus only means that it is to resist all actual and possible forces that may threaten to disorganize it from outside. What it all means is that the organization is to be individuated, i.e. it is to know its own extent and to have an urge to maintain its exclusive nature. Further differentiation of chemical compound therefore means the evolution of consciousness or life in it. A living body has in it a principle of consciousness wherein the extent and organization of the body is reflected. The principle of consciousness directs and guides the body to maintain its integrity and exclusiveness. A living body faces odd circumstances every now and then and as such it is required to struggle with its environment so that it may maintain its integrity and may also grow and develop.

The struggles of a living being may broadly be classified into two categories on the basis of the nature of activities involved in such struggles. These are his struggles with the environment and struggles with his own self. In the first category, a living being struggles with the environment in which he is placed in order to maintain his body and in order to live a happy material life. In the second category, the living being struggles with his own self in order to know it, to enjoy a happy spiritual life and ultimately to be one with the Absolute. After birth, air, water and food are the immediate necessities of a living being to keep intact the organization of its body so that consciousness may continue to exist in it. At the earliest stage of its life a living being of the most primitive form gets air, water and food near at hand and so it need not strive hard. We suppose it to be so because the very evolution of a living body depends on the availability of the said materials in the area. In course of the bodily development and growth of differentiated organs the living being needs various other materials for their upkeep and further developments. The living being then strives to collect the said materials from Nature. This process is more a matter of exploration and collection of food from Nature than a struggle with her. Till the growth of organs for locomotion a living being lives where it was born. After growing organs for locomotion, the living beings move about to collect food, air and water for the body. As according to our hypothesis, the evolution of life in the universe is not an accidental phenomenon but a development aimed at by the Absolute, death due to shortage of food, air and water cannot be accommodated in the plan. So long as a species of living beings live on instincts it may be supposed that Nature herself provides means for their sustenance. Nature of course makes room for evolution of developed species by making extinct some living beings of undeveloped species. Unless a balancing of this nature is done there will surely be shortage of room and food for all. Nature must include in her plan a balancing of this type, if at all She is supposed to have a plan. When due to natural reasons a species of living beings is annihilated we may therefore suppose that it was so intended by Nature. Natural calamities like earthquake, storm, flood etc. may in this sense be supposed to be the means of Nature to balance the evolution of living beings. It may also here be mentioned that the living beings who exercise a freedom in their will also determine to a certain extent the fate of other living beings. Such interference of living beings may of course be considered as their efforts to collect some amenities from Nature. As regards their abode, we have said that at the primitive stage of evolution of life, the living beings have no limbs for locomotion and the very place of their origin is their abode. After the growth of limbs for locomotion, the living beings move from place to place in search of food and for selecting suitable places for their abode. These are activities of living beings that amount to explorations in the domain of Nature. The elaborate ways of living methods developed by the intelligent living beings and their efforts to save their lives from natural calamities are also of this nature. The extreme cases of these explorations are interferences in the courses of Nature. Nature provides means for the sustenance of life and the living beings are supposed to find such means out. If they fail to do it due to improper exercise of their intelligence, Nature cannot be made responsible for it. If no means are provided by Nature for the sustenance of a particular species of living beings, it must be assumed that Nature wishes to annihilate such beings, either totally or partially.

Coming to their own level the living beings enter into a field of competition and start fighting with each other for the share of food, water, air and abode. Even in the case of most primitive organisms there is a competition amongst them for food from the soil and environment. The urge for life forces every organism to collect food etc. from the environment and if the same area is to feed a large number of organisms, the said organisms will have to compete with each other for the purpose. There is a competition of this nature in the kingdom of plants also. Each plant tries to suck its life-blood from the soil and air and struggles with its fellow beings to get its necessary share. In the case of animals who have organs for locomotion this competition and struggle is quite visible. When the quantity of food, shelter etc. are quite enough for a community of living beings there may be a competition amongst them for articles of better quality and suitability. But when the quantity of such articles is not enough for all, there will be a keener competition amongst them and some will perish due to shortage of food etc. Again, one species of animals may be the food for another species. This also means destruction of animals and balancing of life in Nature. The plants convert inorganic compounds into organic compounds and the latter are consumed by some animals as food, because the said animals cannot effect such transformation. Thus, according to the very scheme of Nature some species of living beings serve the purposes of food for other species. The destruction of some species of life for the sustenance of life of other species is thus included in the very scheme of Nature. In the case of lower animals who are guided only by their instincts such struggles and consequent destruction of some of them cannot be considered as interferences in the courses of Nature. The lower animals are guided by their instincts which are nothing but the workings of Nature through them. But it is otherwise in the case of intelligent beings like men. Men do not adjust their lives with the environment. They rather create an environment suitable for their lives by interferences in the courses of Nature and as such they carve their own fate. For most of the events of their lives, therefore, men themselves are responsible.

It may be said in a general way that in the case of living beings who are guided by their instincts the question of 'good' and 'bad' does not arise. They are not actuated by any desire for doing good or for abstaining from any action which is bad. They only move like tools in the hands of Nature. Nature works in them through their instincts and makes them move in the way She likes them to. The question of 'good' or 'bad' therefore does not arise in the activities of their lives. But the ethical question does arise in the activities of those living beings who exercise their freedom of the will and employ their intelligence. They make themselves responsible for their activities.

The summum bonum of life being its ultimate union with the Absolute, the activities of a living being are good for him when these are conducive to his progress towards such union and bad when his activities drive him away from such union. An intelligent individual is thus burdened with a heavy responsibility, the responsibility of making his activities good, and to avoid meticulously the activities that are bad for him. Every step in the life of an intelligent being needs thus to be based on a well calculated plan. He is to keep his intelligence heavily engaged during the whole tenure of his life. There is no moral holiday for him.

Coming again to the question of struggles of living beings amongst themselves we find them competing with each other for suitable mates for procreation. It is based on their urge to continue to live for ever as also to enjoy carnal pleasures that excel all pleasures of such kind. Competition and struggle on this account take sometimes an extreme form resulting in the death for some of them. ·Next to this is the struggle and competition for land. The plants and lower animals fight with each other for land mainly for the purpose of collecting food. The beasts prefer to have particular areas of land for hunting purposes and fight with each other on this account. In the case of primitive men there is a similar fight either with each other or amongst various communities for possession of some particular areas of land for the purpose of hunting animals, for drinking-water, for enjoying fruits etc. or for cultivation and production of grains etc. Wars are also common between civilized communities for the possession of land for minerals, for strategic positions and for similar purposes. Then comes the conflicts between intelligent and civilized beings for the acquisition of wealth, collection of amenities for comfort and provisions for future generations.

The living beings also fight with each other for the enjoyment of power and authority over others. This is due to inequality in their strength, physical or otherwise. Some member or members of a species may become stronger and more powerful than others. The weaker ones will then obey the command of the stronger ones for the security of their lives and for the safeguard of their properties, if any. The leaders or powerful members of a species enjoy an inexplicable satisfaction in their acts of commanding over others and making their commands obeyed. There is perhaps no struggle on this account in the plant kingdom and in the life of non-intelligent animals. But it is quite common amongst the intelligent beings, particularly amongst men. As the leaders of societies have a privileged position they are always to face their competitors who wish to have the same privileged position, and its invariable result is struggle and conflict in some form or other. Intelligence and civilisation may make these struggles sophisticated but struggles on this account continue to exist at different levels of intelligent life.

Intelligent beings like men also fight with each other to keep intact or to regain their esteem. The sentiment of keeping one's esteem is a sign of the development of one's awareness of his own self and individuality. When a man is intelligent and matured enough to be aware that he is an individual amongst his other fellow beings and that he has a position amongst them, he develops a sense of esteem. He does not like to have his position lowered in his community or society. Development of this sense of esteem is possible for those intelligent beings only who are aware of their individualities, who have a clear apprehension of themselves as having exclusive lives of their own. Development of the sense of esteem is an early stage of the development of self-consciousness. The plants and lower animals cannot have any sense of esteem because they are not conscious of their own selves. Sense of esteem is a sentiment that plays a role in the life of human beings and it sometimes dominates and determines their behaviour. Men fight with each other to gain or to regain their esteem. Men not only desire to have a position amongst their fellow beings, they also want their families to have a position amongst other families, their societies amongst other societies and their nation amongst other nations. Any loss or lowering of their position incite them to fight with others in order to gain or to regain their personal esteem, their family esteem, their social esteem or their national esteem.

In course of living their lives the intelligent beings like men attain a still higher stage of the development of their individualities. They not only become conscious of their own lives but also of life as such. They formulate principles and ideals for easier and happier life. They uphold socio-political ideals and may also develop personal ideosyncrasies. They try to follow the said principles so as to attain their individual and socio-political ideals. Whenever conflicts arise or forces stand in the way they try to fight it out. While facing their environment and facing conflicts and struggles in life men become conscious of a vastly superior power which dominates over all. They develop an attitude of respect and reverence towards that power and try to appease its anger and to gain its favour so that they may live their lives peacefully and enjoyably or may enjoy a happy existence for ever. Thus there originates a religious consciousness in them. They perform certain acts to express their religious feelings and thus appear the religious ideals in their lives which they try to uphold at any cost. There may be conflicts and struggles between various communities of men who uphold different religious ideals with widely different dogmas, rituals and ceremonies. Individuated and intelligent beings like men thus fight with each other on account of personal ideosyncrasies or ideals and on account of their socio-political and religious ideals.

Further development of his individuality makes an intelligent being like man more conscious of his own self. He goes to explore the mysteries of this universe in order to determine his own position in it and to know the significance of his life and its destiny. In course of analyzing his knowledge of the external world a man ultimately comes to know the real nature of his self and the summum bonum of his life. He comes to know the limitations of scientific knowledge and also the fact that scientific analysis cannot help him to know the mysteries behind this universe. The sciences cannot also determine the value and significance of his life. A man therefore enters into the domain of metaphysics and religion. Metaphysical analysis leads him to conclude that his ultimate destiny is to be one with the Absolute. He also comes to know that in order to attain his ultimate goal he is required to explore and master the workings of his own soul. He therefore enters into the field of his hardest struggle and it is with himself. In order to master the technique of acquiring true knowledge of his own self a man faces various obstacles. He finds that his mind is swayed constantly by his desires, emotions and passions. His love for power, lust and earthly pleasures make his mind extrovert and so he is to struggle hard with himself and to direct his mind towards itself. While discovering the real nature of his mind and its principles, the man faces troubles due to obstructions, anomalies and puzzles. Acquiring knowledge of his own self and also thereby acquiring knowledge of the universe (which is reflected in his self) the man ultimately reaches the stage of having infinite knowledge in his consciousness and becomes one with the Absolute.

Now, the struggles of living beings on various accounts, as enumerated above, do at the same time retard or accelerate their progress towards the ultimate goal. In the case of creatures who are guided by their instincts only, the conflicts and struggles they are in are for the sake of sustenance of their dear lives only. The principle of individuation in them is only at the stage of half-way development. They live only in their present and cannot conceive of their future. All their activities are confined in the sphere of collection of food, occupation of places for shelter, procreation and protection of life from dangers. These conflicts and struggles are inevitable in their lives because due to lack of intelligence they cannot formulate or adopt principles to ease the tension on this account. Ethical judgments cannot be passed over their activities. They only feel the urge for life during their tenure of existence in a physical body and die to be born in a higher form of life. Conflicts and struggles on account of food, shelter, procreation and protection of life from dangers are also inevitable in the life of intelligent and self-conscious beings like men. But men not only live in their present but also calculate and foresee their future. By exercising their intelligence therefore they formulate and adopt principles to ease the tension in their lives on this account. Conflicts and struggles on account of the acquisition of wealth and earning prestige are also almost inevitable in the life of men. Thoughts for future provisions of food and other amenities make men strive to amass wealth. The development of individuality makes a man conscious of himself as a person amongst others. He also becomes conscious of his position and status in his family, in the society to which he belongs and in the nation or tribe of which he is a member. A man always strives to have a better status for him, for his family, for his society and for his nation. Conflicts and struggles on this account are therefore matters in the natural course of events in a man's life. But conflicts and struggles that come in the way of gaining control over others, of fulfilling the socio-political and religious ideals and of gaining control over one's own self are not inevitable phenomena in a man's life. A man may not have any fascination for enjoyment of power and control over others. There will not be any conflict or struggle on this account in the life of a man of such nature. Conflicts on account of socio-political ideals also are not inevitable in the life of all intelligent beings. Even without any entanglement in such conflicts and struggles, an intelligent being can attain the summum bonum of his life. These conflicts are concerns for individuals who crave for power. But though material prosperity in life depends to a large extent on these conflicts and struggles and though the material aspect of life must be made conducive to the real development and progress of a man, an aspirant for union with the Absolute hardly does consider these conflicts and struggles as his concern. He practically renounces the earthly pleasures and as such does not like to be troubled with these conflicts and struggles. Religious ideals in the true sense of the term are matters concerning the inner life of an individual and so conflicts and struggles within himself are inevitable in the life of an individual who wants to fulfil the mission of his life. According to us, in the normal course of events an intelligent being will have to face the ups and downs of spiritual life sooner or later.

As products of Nature, all living beings attain the goal of their lives when they tune up their activities with the rhythm and courses of Nature. In other words, living beings are supposed to follow Nature. The creatures who are guided by their instincts only, who do not exercise their will, are not deviated from the courses of Nature. They are, in this sense, guided by Nature Herself. But in the case of intelligent beings, it is otherwise. The intelligent beings have freedom in the exercise of their will. They apply intelligence, act according to their desires and carve their future themselves. Their thoughts and activities may therefore be good or bad, conducive to the attainment of their ultimate goal in life, or not. Even though an intelligent being acts according to the dictum of his own will he need not be away from the courses of Nature. He may tune up his activities with the rhythm of Nature, if only he desires to do so. In course of living his life, an intelligent being observes that some of his acts bring happiness for him and some only miseries. He therefore brands some such acts as good and some bad for him. He tries to perform the good acts and avoids the bad ones. He formulates certain codes of behaviour, tries to be habituated in obeying them and strives to derive the maximum possible benefits from his activities. He also tries to perform the acts efficiently and profitably. Thus an ethics of utility evolves automatically in the life of intelligent beings. Intelligent beings observe that their activities in connection with the collection of food and protection of life and properties from dangers can be more effective if they live not alone but together in a group with others. They therefore live in families along with its other members. In higher stages of the development of their intelligence they also live in communities, in societies and in States, because a life lived in these conditions are more beneficial and enjoyable than that of a life lived alone. An intelligent being also observes that the pleasures of his life depend to a large extent on the help he gets from others. In order to make others to help him, he himself therefore helps others in many possible ways. These are his acts based on an ethics of utility in the sense that these are centred only in his self-love. The urge for life is present in every living being. Every such being wants to live and to have the joy of living. To make this possible, an intelligent being obeys the codes of this ethics of utility. One who defies the codes of this ethics is likely to suffer while collecting his food and protecting his life and property. But even if an individual abides by these codes of utility and acts accordingly, he may not be in tune with the rhythm and courses of Nature. It may be so due to the fact that the codes of this ethics are formulated on the basis of benefits that individuals of a particular species may derive through certain activities. Some or all of these activities may be beneficial for a particular species of creatures or some creatures of a particular species but harmful to creatures of other species or to other creatures of the same species. For example, by selfish activities an individual or a group of individuals may derive benefits for themselves only but at the cost of others. It must be said that the individuals act in this manner when they think in terms of themselves only and not in terms of other creatures and objects of the universe. Their viewpoint is thus partial and one-sided. As all objects and creatures are at the bosom of Nature, She cannot act like selfish individuals. Thus the selfish activities of individuals will be out of tune with the courses and rhythm of Nature. The ethics of utility may not thus bring benefits of ultimate value for the individuals. As the ultimate goal of life is its union with the Infinite, only those acts are of supreme value which accelerate the progress of an individual towards that union. The ethics of utility may not help in this regard, unless it is in tune with the rhythm of Nature.

As per our hypothesis, the course of Nature is an evolution of matter and life and their ultimate absorption in the Absolute. To determine the course of Nature and to make one's will conform to it is thus the real ethics for an individual. It does not of course mean that the ethics of utility is necessarily antagonistic to the real ethics for an individual. It does only mean that an individual is supposed to affiliate his ethics of utility to the real ethics for him. Is it possible for an individual to defy the ethics of utility altogether? According to us, it is impossible, because an individual can never avoid the activities for collection of food etc. and protection of his life from dangers. An individual must therefore have an ethics of utility in some form or the other, though this ethics does in due course reveal its limitations and one-sidedness. The individual observes in due course that he cannot attain the summum bonum of his life by following only the codes of such an ethics. He therefore tries to universalise the codes of his ethics of utility or to affiliate it to the universal laws of Nature. This is nothing but to modify this ethics so that it may conform to the courses and rhythm of Nature.

A little reflection on the courses and movements of Nature will convince an individual that the evolution of entities in Nature is directed towards the manifestation of life and mind. We make this observation in view of the following facts. All movements in inert matter ultimately produce only compounds of them, either inorganic or organic. The transformation of inorganic elements into inorganic compounds is nothing but a repetition of the process in the sense that the basic nature of the constituents and the product do remain the same. But in the case of a transformation of inorganic elements into an organic compound, the said compound acquires new characteristics and, at the stage of a particular organization in it, manifests life. It may here be asked: how can there be life and mind in the stars and planets which have radiated heat and are too cold to evolve and support life on them. We shall here give a very simple answer. Explosion in such stars and planets can transform their chemical composition releasing the necessary heat to evolve and support life on them in due course. We are of the opinion that the very aim of Nature in evolving entities in the universe is to manifest life and mind and the summum bonum for life is its union with the Absolute. The scheme of Nature is therefore to promote a general welfare of the living beings. It is of course a fact that Nature balances life by causing death for some beings. But this balancing of life is also a part and parcel of the scheme of general welfare of living creatures and their progress towards union with the Absolute. To be in the courses of Nature, therefore, one is to look after the welfare of life in general and to help the living creatures in their journey towards the Absolute. To promote life and to look after the general welfare of all is thus, according to us, the first principle of real ethics for an individual. The codes of his ethics of utility should also be formulated keeping in view this first principle of real ethics. In fact, the codes of the ethics of utility should be formulated in such a way that they conform to this first principle of real ethics. The codes of ethics should be such that they are beneficial not to a particular individual or to a group of individuals but to all creatures in general. It means that the codes of ethics should be universal principles for the general welfare of all. To formulate principles of intelligent behaviour, the most practicable corollary that can be derived from universal ethics will be: 'to treat others as if they are one's own selves'. An individual's behaviour towards others should be just what he expects from others towards him.

The activities of an individual will conform to the courses and rhythm of Nature only when he acts like her agent. The individual is to act in such a way that the purposes of Nature are served thereby. He is to help Nature rather than to obstruct Her courses. He is to formulate his codes of ethics by drawing lessons from Nature. He who looks after Nature will be looked after by Her. If, instead, an individual defies all ethics, he will have to face obstructions all through his life. If he has no ethics of utility, he will be unable to feed himself regularly. He will not be able to settle anywhere permanently and will have to face antagonistic forces constantly. This position is of course impossible for a man because a man cannot but have an ethics of utility. If an individual's ethics of utility is totally out of tune with the courses and rhythm of Nature, he may at best be able to maintain his body and live a life of sensual pleasures. Being on a wrong track, the growth and development of his mind is sure to be affected. His knowledge of the world around himself will be wrong. Being guided by his wrong ideas he will drive himself away and away from the Absolute who is perfect knowledge itself. As the said individual will be away from the courses and rhythm of Nature, his efforts for attaining the state of an ever-happy existence will not be backed by Nature. Nature will rather put hindrances before him. All efforts of an individual on the wrong track will thus be futile and, instead of attaining an ever-happy existence, he will only be engulfed in doom.

Chapter 8
Knowledge and its Origin

In course of developments due to movements in the corpuscles of energy, compounds of energy, or elements are formed. When a particular mass of such compounds develops the principle of exclusion in it, i.e., when it develops the capacity of retaining itself as an entity separated from other entities and also the capacity of drawing in materials from outside for conversion to compounds identical with its own kind, consciousness is manifested in it and it becomes an individual living being. We have shown earlier that consciousness is not an element foreign to matter. In fact, matter implies consciousness and consciousness matter. Consciousness was there in matter, though in an implicit form and after attaining the principle of exclusion the matter in question manifests that consciousness. At the very first moment when consciousness is manifested in the compound of elements, the said consciousness can have only that compound as its content, including of course the state of that consciousness itself. There cannot be anything else as the content of such consciousness. After the manifestation of consciousness in a particular compound of elements, it becomes limited by the compound and also qualified by it, for, otherwise, there will not be any difference between consciousness manifested in two different bulks of such compounds. Though consciousness as such is an indivisible entity covering the whole universe, at its manifestation in a particular compound of elements it becomes limited by the said compound. It is now clouded by the bulk of energy which is reflected in it. Consciousness is unlimited and infinite when it covers all entities of the universe and limited when it covers only a particular bulk of energy. If it is not so, then there cannot be any reason why should consciousness appear in a particular bulk of energy. From the standpoint of energy also we find that when a compound of elements develops the principle of exclusion in it and is able to maintain its integrity it needs a principle of intelligence for its being guided by it. Consciousness is the said principle and its sole function is to guide the activities of the compound of elements in which it has appeared. Consciousness guides the compound of elements so that it may adjust itself with the environment and may meet its nutritional needs. If the consciousness, manifested in the body, is infinite then the body will not have the principle of exclusion in it because it will not then be an entity separated from other entities. But this, according to our hypothesis, is not the case. The bulk of individuated energy needs consciousness to make further progress in its movements. The primary concern of consciousness is therefore the progress of the said bulk of energy. For the development of this bulk of energy only, consciousness goes on acquiring knowledge of other entities in the environment, or, rather goes on covering other entities till it covers all entities of the universe and regains its original state of infinitude. Thus, from the standpoint of energy and its individuation, consciousness, manifested in a particular body, is limited and qualified by the body. If infinite consciousness is manifested in a bulk of energy at the very beginning then there will be no reason why the body should commit errors and mistakes in its activities. Thus, from this standpoint also it will be seen that consciousness, manifested in a body, is limited and qualified by the body.

There is a mind in consciousness, manifested in a compound of energy, which acts as its centre. Can it be otherwise? Can there be consciousness, manifested in a body, without a mind? We do not think so. Consciousness evolves in a body to guide it in its activities for maintenance of its integrity and for drawing in materials from outside for nutritional purposes. The main content of consciousness is therefore the body in which it has been manifested. Consciousness must have in it a reflection of the body and its requirements. And, to direct the body to act in a particular way, consciousness must have in it a reflection of the environment in which the body is placed. The reflection of the body and its environment are analysed in consciousness and a line of action is determined for the body so that the body may meet its requirements. Now, the body and its environment are not of the nature of consciousness, because, while the former are material principles, the latter is spiritual. An entity, matter or consciousness, cannot enter directly into the apprehensive principle of consciousness manifested in a body. An entity from outside can have its reflection in the said consciousness only after its conversion into an apprehensive principle. For the conversion of outside entities into an apprehensive principle there must be a centre in consciousness and this centre, according to us, is the mind. Entities from outside enter into one's consciousness through his mind. The said entities first take the form of ideas in the mind. The ideas are analysed in the mind, are related to other ideas already existing there, and then integrated with them. An apprehensive principle is thus formed and accommodated in consciousness. Even though mind has a distinct function of its own, it is not anything other than consciousness, because, consciousness is a mental entity. Mind and consciousness are only two strata of the same entity. This will be clear if we analyse the position from the standpoint of memory etc. Experiences of the past events remain in consciousness as memory though the same are not present in the mind. Thus entities in the apprehensive form do exist in the subconscious and unconscious regions of the mental principle of a living being. This shows that there are different regions in consciousness and all the regions are not equally active. For the sake of our analysis we differentiate consciousness into two regions—consciousness as a whole, and its centre, the mind, where materials from outside are first converted into the mental form and then absorbed in consciousness.

We have said earlier that due to necessities of life in the organism, consciousness directs the movements of the body towards formation of sense-organs which are to collect data of information about the environment. After collecting such data of information the sense-organs pass them on to the mind which analyses them, integrates them with other data collected earlier and forms apprehensive principles to be absorbed in the consciousness concerned. After appraisal of the situation consciousness sends through the mind directions to the body for its movements in a particular line. In every knowledge situation therefore the entities involved are: consciousness as the subject of knowledge, mind of the subject, sense-organs of the subject and sense-data from the environment, or the object of knowledge which includes both energy and consciousness in other organisms. The sense-organs of a living being come in touch with entities in the environment or rather the sense-organs are affected by such entities. These affections in the sense-organs are then transmitted to the mind. The mind analyses these affections, or impressions, and converts them into ideas. What, then, are these ideas? An idea is not the affection in the sense-organs because the affection is physical whereas an idea is mental. Though an idea represents the affection in a sense-organ it itself is not the affection. An idea is therefore a symbol in the mind that represents the affection in the sense-organs. Can a single idea form a part of consciousness? According to us, it can, but only as a change in consciousness. A single idea can only make consciousness alert over a situation. An alert consciousness waits for further developments in this regard. When more sense-data of the environment are collected by the sense-organs and passed on to the mind the first idea is integrated with them and an apprehensive principle is formed. This apprehensive principle is analysed and studied with reference to the requirements of the body and a line of action for the body is formulated. In the case of forming apprehensive principles concerning the conditions of the body, the sense-organs may or may not have any role to play. For example, in the case of an outward injury in the body the sense-organs will collect sense-data from it and apprehensive principles will, as usual, be formed in the mind. The mind in such a case will direct the body to act in a particular way. But in the case of hunger and thirst the sense-organs transmit nothing to the mind. In the case of activities due to promptings of the so-called sixth sense, the five sense-organs do not function at all. In the case of an unsound stomach the sense-organs do not transmit any sense-data to the mind but the mind becomes aware of it and ideas are generated therein. All these cases show that consciousness comes in touch with the material objects sometimes through the medium of sense-organs and sometimes without them. This position also shows that there are functional differences between the mind and consciousness, though basically they are the same. We learn from it further that consciousness tries to know the condition of the body and its environment and guides the body to move in particular ways to maintain its principle of exclusion or individuality.

The ideas are only symbols through which the mind and consciousness of a living being apprehend the outside entities, including also the body of the said being. After all, the ideas in the mind are only the representations of entities, they are not the entities themselves. As such, the ideas cannot but be considered as symbols of entities. The whole edifice of knowledge and apprehension is based on these symbols. As principles of apprehension, these symbols are mental. As principles of expression these symbols represent entities in the universe.

Being ultimately based on affections in the mind of an individual, knowledge is always the knowledge of a particular individual, and universal knowledge in this sense is only a myth. Whenever two individuals communicate with each other by using ideas of entities, they will mean different things and as such the said communication will be impossible. But such is not the case, actually. Universality in knowledge is possible because the forms of understanding and expression are universal in character. There are certain forms of understanding which are universally true for a particular species of living beings. There may of course be differences in the forms of understanding and expression of living beings belonging to different species. The reptiles and carnivores may have edifices of knowledge quite different from that of ours. The knowledge of lower animals may serve the purposes of their lives but from the standpoint of truth our knowledge is nearer to it than those of them, because our knowledge is not the knowledge of particular entities only. Ours is the knowledge of all entities of the universe and the edifice of our knowledge is based on the coherence between them. Our knowledge is thus rooted on a stronger base and so is nearer to truth.

Coming again to the question of universality in knowledge our assertion is that we apprehend entities through certain forms of understanding universally true for all of us. Whenever we go to understand each other, the said forms of understanding act like bridges between us. These forms are our laws of thought and the concepts of space, time and causality. The laws of thought are principles of understanding and they refer to states in consciousness. The laws of thought are laws of consciousness. They reveal to some extent the nature of consciousness. The concepts of space-time-causality are related to energy, or rather to the understanding of configurations of energy. Though these concepts are mental, they refer also to the evolution of energy in the universe. All other principles that work in our faculties of understanding are only subordinate to the universal laws of thought and the concepts of space, time and causality. These universal principles being present in our minds, inter-communication of ideas between us is possible. Whenever we express our ideas we express them in the framework of these universal laws which are comprehensible to others. When we express anything relating to mind and consciousness we follow the laws of thought which are also followed by others. We thus can understand each other easily. Whenever we express anything relating to energy we do it in the framework of space-time-causality which others also do. Our expressions relating to configurations of energy can therefore be understood by others. In view of the above it is quite possible for us to build a mass of universal knowledge.

Let us now see whether the ideas acting as symbols of affections in the mind are authentic and if so, how. According to us, every idea formed in the mind may not be authentic in the sense that they correctly represent entities outside the mind. In order to be authentic, every idea must stand two tests—the first being that the acts based on the idea must show that it represented the concerned entity or entities correctly and brought in expected results and the second being that the idea must be coherent with other authentic ideas already existing in the mind. The authenticity of ideas depends on the one hand on the sense-organs, nerves etc. of the individual and on proper analysis and understanding on the other. If the sense-organs and nerves etc. are in proper order and they function in the proper way and if the mind analyses the affections in the sense-organs properly then the ideas formed in the mind are likely to be authentic.

Now, what is the difference between affections in the sense-organs and the ideas formed in the mind to represent them? The difference between them is that the affections are centred in the body but the ideas are centred in the mind. If a sense-organ of the body is paralysed or if it is removed from the body then a particular type of affections will be absent in the body and their corresponding ideas will not be formed in the mind. It shows that the affections are centred in the body and ideas in the mind. If the affections in the sense-organs are not analysed properly, as are done by children and idiots, then the ideas formed in the mind will not be authentic in the sense that the ideas will not represent the outside entities correctly. Now, symbols or ideas in the mind are representations of the affections in the body. Are the ideas also affections in the mind? In a sense, they are, because the very process of formation of ideas affects the mind. There can be affections in the mind only when ideas are formed in it. The mind comes in touch with the entities outside through the medium of sense-organs and as such the ideas are only representations of those entities. But as affections in the mind they are genuine for the mind, though they may not be true as matters of fact. To a mind its ideas are real because they are based on real affections in it. In the case of ideas that do not ultimately stand the test of coherence and correspondence, the mind modifies them later. But till modification, the ideas are quite real for the mind and the mind directs the body to act in accordance with these ideas.

Can the mind come in direct touch with the external world? Can there be a connection between mind and the external world without through the medium of sense-organs? The mind cannot ordinarily come in touch with the external world, except with the body in which it has been manifested. According to our hypothesis, consciousness and mind evolve in a body to complete its movement towards individuation and to guide its movements. Consciousness and mind are therefore related to the body. All events in the body will therefore have reflections in its consciousness. In the case of an organism who has not yet grown and developed sense-organs, consciousness itself will make it work instinctively. Such an organism has no brain and mind to deliberate on matters. Consciousness works in it directly and makes it to act instinctively for the purposes of life. Its acts are instinctive, because, it has no brain and mind and as such it cannot exercise its will in the deliberations involved. An organism of this type has no organ to store up its experiences and cannot also know beforehand the possible results of its acts. It has no scheme for its future activities. It practically acts as a machine steered by its consciousness. An organism which has evolved a brain has at least a mechanism to store up its experiences. It has also a mind to deliberate on new situations. In case its brain remains undeveloped and its functioning is poor it cannot exercise its will to determine a line of action in novel situations. Its activities are thus more instinctive than rational. But in the case of a living being whose brain is developed, there is a mechanism for storage of his past experiences. It can therefore link up new situations with his past experiences. It can think about the matter properly and by exercising his will can make a plan of his future action. In its case therefore a large number of activities are controlled and guided by its mind. Information about conditions of different parts of the body comes to its brain through the nerves for analysis and deliberation. We therefore draw here the conclusion that an organism which has grown a brain exercises his will in mental activities. This development in an organism is a development towards individuation.

In the case of an organism which has no brain, deliberations are done in consciousness and the organism has no will of its own. But in the case of an organism which has a brain, some of the deliberations are done in its mind and the organism exercises its will. It may therefore be concluded that the functioning of the brain and the mind and exercise of the will are simultaneous. Now, mind, according to us, is a centre in consciousness where deliberations on matters are done. Even in organisms who have no brain and who do not exercise their will there are deliberations in their consciousness. Mind, in our sense of the term, is, therefore, present in the consciousness of such organisms also. The existence of a brain and exercise of one's will are thus correlated to each other. Various items of information come to the brain through the sense-organs and the nerves. These are stored up in the brain in the form of memory. When deliberations are done taking the elements of memory into account the organism exercises its will consciously. Evolution of memory is thus the basis for exercise of the individual will.

Coming again to our point at issue we say emphatically that consciousness and mind, being centred in a particular body, cannot, generally, come in direct touch with anything other than that body. Direct communication with anything other than the body is possible only when the mind crosses the barrier of the body and attains infinitude. It is so because in such conditions the mind becomes one with infinite consciousness, wherein is reflected all entities of the universe. At this stage an individual mind comes in touch with every other entity of the universe even without the medium of sense-organs. This is the ultimate stage that an individual may attain through evolution. Even before attaining this last stage of evolution, an individual may come in direct touch with other entities through intuition. It is an individual's intellectual sympathy for or oneness with the object he knows. But it is only a sudden and temporary phenomenon. When one's intellectual apprehension of an entity brings him nearest to the truth of that entity, or when the whole truth of that entity is apprehended by him, he becomes one with that entity. There is at that stage no difference between his apprehension and the truth itself. The body and sense-organs of the individual concerned play no part at that state. Intellectual oneness with the entity reveals also other intellectual facts about the entity, because, there being no difference between the intellect concerned and the truth of the entity, nothing intellectual can remain unreflected in the consciousness of the individual. But this intuitive knowledge of an individual is the knowledge of a particular entity only, not of all entities of the universe. Such knowledge is therefore incomplete and is not based on perfect truth about the entity. Moreover, due to its preoccupation with the body, the mind comes down to the ordinary level of existence soon. The state of intuitive apprehension is thus only a fleeting phenomenon in the total life of the individual. It may here be mentioned that in the ultimate analysis such intuitive knowledge will be found to be based on some sense-data of the individual concerned.

As regards knowledge of oneself, an individual is supposed to undertake a journey within. To acquire knowledge of itself the mind is required to discover the laws and processes of its own movements. The laws and processes of movements of the mind of an individual are the same laws and processes that lie inherent in the infinite consciousness of the Absolute. But an individual's consciousness is centred in a body and as such his mind is ever engaged in the affairs of the body. The bodily desires and passions keep the mind always engaged in these affairs. The said desires, passions etc. act like anti-forces against the mind's attempts to concentrate on itself. They always drag the mind down to their own level. We do not, of course, mean here that the mind could free itself from the affairs of the body had it not been so dragged down. According to our hypothesis, the very purpose of the evolution of consciousness and mind is to guide the movements of the body enabling it to sustain its life and to develop its individuality. The mind cannot therefore shirk off its responsibility towards the body. It can however confine its activities to the barest minimum of bodily requirements. In doing so, the mind can save time for its journey within and can discover its inherent laws and processes. It is not an easy task for a person to withdraw his mind from his desires and passions. Certain types of discipline of the mind do however make it possible.

The mind comes in touch with configurations of energy in the body and, through the body, with configurations of energy in the external world. In course of the analysis of affections in it due to energy, the mind goes on discovering the laws of energy which are the so-called laws of science. The laws of science are general and they cover all the particular bits of energy that exist in the universe. Even though an individual comes in touch with a very small fraction of the total energy manifested in the universe, his knowledge of the general laws of science makes him aware of truths underlying all such energy. But this knowledge of an individual is knowledge of the general nature of energy only, not the knowledge of their specific characteristics. It is so because an individual does not come in touch with the total being of those bits of energy unless he becomes one with them. Specific and exact knowledge of all manifested energy can be obtained by a person only when he has attained the state of infinite consciousness.

What role does the scientific knowledge of a man play in his effort to attain infinite consciousness or to be one with the Absolute? According to us, such knowledge of a man plays in this regard a very insignificant role. Consciousness is the depository of all knowledge and it is evolved in a body to guide it in its activities for sustenance of its life and for its movements towards individuation. The body requires materials for its sustenance and so it draws in materials from the environment. In course of doing so, the sense-organs collect information about the environment and transmit such information to the consciousness concerned. Consciousness analyses the data of information, gathers comprehensive knowledge of the environment and directs the body to move in some particular ways. All knowledge of the environment are ultimately brought under some scientific formula. As regards the concerned body, the mind comes to know its conditions and workings either through the medium of sense-organs or without them. And all knowledge of the body can also be brought under some scientific formula. For an individual the only difference between his body and the environment is that the mind acquires knowledge of the body both directly and through the sense-organs but acquires knowledge of the environment through the sense-organs only. The body has this privileged relation with the mind because the mind has evolved in it. But in spite of this privileged status of the body, it is outside the area of mind like other objects in the environment. Whatever idea there is in the mind, either relating to the body or relating to the environment, it can only be reflections of some object or objects and these reflections are analysed and compounded by the mind either correctly or incorrectly.

The scientific knowledge of the body and of other objects do never make the mind one with them because such knowledge is knowledge only of the behaviour of energy contained in the environmental objects and in the body and not of the knowledge of such energy as they are. Moreover, a man's knowledge of the environment and some parts of his body are the products of his analysis and synthesis of sense-data received through sense-organs. This knowledge is therefore relative and indirect. It is relative because it depends on the sense-organs. It is indirect because it is based on sense-data which only represent some objects and are not those objects themselves. This knowledge is thus inferential and constructed out of the data through sense-organs. But it is otherwise in the case of direct apprehension of the bodily conditions. In this case the bodily condition is apprehended first and then it is analysed and arranged in the form of thought. In the former case apprehension was posterior to the process of thought but in the latter, apprehension is prior to the process of thought. Direct apprehension is made clear in thought where the different elements of apprehension are analysed and separated from each other. Any defect in the analysis may be corrected by comparing it with its original source, i.e. with the apprehension in its totality. But in the latter case thought is constructed earlier and it cannot be compared with the original source. Only the workability of objects constructed in such thought can determine whether it was correct or not. If apprehension of the form of a mass of integrated objects is the final stage of human knowledge, then direct apprehension is superior in status to indirect apprehension through thought, as is the case in scientific knowledge. Scientific knowledge is indirect and relative and it only serves the purposes of our day-to-day life. But, for real knowledge of the extra-mental world, a feeling of oneness with it is necessary. This feeling of oneness is absent in scientific knowledge because it is based on elements of thought. As this feeling of oneness is present in direct apprehension, it is superior to scientific knowledge as a means to attain infinite consciousness.

Scientific knowledge only provides us with information about the external world and the world of mind and information of such nature can only be utilized for the practical purposes of life. By utilizing scientific knowledge we can make the sustenance of our life easy and may also have material enjoyments but for acquiring real knowledge of the external world a feeling of oneness with it, or direct apprehension, is necessary. And for this, one must enter into the depth of one's own mind and consciousness so as to discover its laws and processes and to develop the technique of feeling oneself identical with the objects in the extra-mental world. Scientific knowledge and its utilization for material benefits divert a man's attention from the path of reaching his ultimate destiny because involvement in the material world obstructs a mind's journey within. It does not of course mean that one can continue to live even without any scientific knowledge of the external world. The best course for a man is to minimize the requirements of his bodily life and enjoyments so that he may have time enough to undertake his journey within.

Life evolves in a body to complete its movements towards individuation. It guides the body in its movements for retention of its principle of individuation. All that happens in the body are reflected in consciousness or life. Due to a very intimate relation between the body and the mind, the latter considers the bodily conditions as its own and confuses itself with the body. The directives that are transmitted by the mind to the body take therefore the form of desires and all that the body fails to do remain in the mind as unfulfilled desires. Being deeply absorbed in the affairs of the body, the mind retains the impressions of these unfulfilled desires and tries after death to be manifested in another body to fulfil those desires. But in due course, the mind realizes that its attachment with the body is not conducive to its attainment of the ultimate goal. The mind comes to know that it can attain infinite knowledge only by exploring its own laws and processes and by making itself one with the objects and entities of the universe. When the mind is able to realize its own nature it crosses the barrier of the body and gradually detaches itself from the body so as to absorb all entities of the universe and to be one with the Absolute.

Though consciousness, by its very nature, is infinite, when it is manifested in a body, it becomes finite. It is so because consciousness is now confined in a body and its infinitude is clouded by the limitations of the body. A body by its very nature is limited because it is an embodiment of the principle of individuation. Being intimately connected with the body, the mind confuses itself with the body and limits itself with the desires and will for fulfilment of the bodily requirements. But when the mind crosses the barrier of individuation and makes itself free from all desires and passions, it goes on enlarging itself and ultimately acquires infinite knowledge or goes back to its original abode, the Infinite. Except as a means to fulfil the requirements of the body, the so-called scientific knowledge of a man has no role in this process of evolution of consciousness in his body and its ultimate absorption in the Absolute.

Affections of every kind in the mind are real in the sense that they are real affections. Affections in the sense-organs are also real in the same sense. But the knowledge of external world as may be constructed by the mind by compounding the affections in it may not be true. The compounding of affections in the mind depends on the conditions of the mind, i.e. on the level it has attained in course of its self-analysis and consequent mental development, on its knowledge of the external world and on their correctness or incorrectness.

The principles and processes of workings of the mind determine the process of compounding of the affections in the mind. If the principles and processes of the mind are fully known and the compounding of affections are done correctly then only one's knowledge of the external world may be correct. Inadequate knowledge of the said processes must make one's knowledge of the external world incomplete. The affections in the mind are caused by the affections in the sense-organs which are transmitted to relevant areas in the brain through the nerves. If the relevant area of the brain, or the nerves, or the sense-organs that come in touch with the external world, do not function properly then the affections in the mind will be incorrect representations of the external world. The affections in the mind thus also depend on the body. Knowledge of the external world is thus relative because it depends on the sense-organs, the nerves, the brain, the mass of knowledge acquired previously and one's knowledge of the working processes of the mind.

Affections in the sense-organs are caused by the data of sense coming from the external world. These data must be real because otherwise the affections in the sense-organs could not be real. It does not of course mean that the constructions out of these data do always represent the external world correctly. In dreams, hallucinations, reveries etc. the data of sense are compounded incorrectly into unreal objects in the external world. The objects constructed in dreams etc. do not represent the external world, because they do not work according to expectation. Workability, according to us, is the main criterion of truth for any object constructed by the mind. Workability of the constructed object proves whether the perceptions correspond to the objects perceived. In those cases where the criterion of workability is not applicable, the criterion of truth will be coherence of the objects constructed with all other known and workable facts of knowledge. The criterion of workability may not in many cases be verifiable from the human level but it does not mean that the objects constructed in the said cases do not work. They work in the scheme of Nature all the same though they may not be of any consequence to human life. Thus, as criterion of truth, workability, correspondence and coherence are interconnected with each other. If an object constructed by the mind is workable, it must correspond to the object in the extra-mental world and it must also cohere with all the known and workable facts of knowledge. If an object constructed by the mind corresponds to the object in the extra-mental world then it must be workable and it must cohere with all the known and workable facts of knowledge. Similarly, if the object constructed by the mind coheres with all the known and workable facts of knowledge then it must be workable and it must also correspond to the object in the extra-mental world.

Now, what actually the sense-data are? Are they mental entities? No, because mental entities cannot affect the sense-organs. A physical sense-organ cannot be affected by anything non-physical. A physical sense-organ is affected by a mind only when that mind is included in the exclusive principle of the body, i.e. when the said mind has been manifested in the body concerned. Having been manifested in the body, a mind has an intimate relation with it and they react on each other. But being physical a sense-organ cannot be affected by any other mental entity because that entity is included only in the exclusive principle of another physical body. Mind evolves in a physical body to complete its principle of individuation. As such, any other mind, evolved elsewhere, cannot affect it. But a physical sense-organ is always lying in contact with other physical entities and so there is interaction between them.

In view of the above, the sense-data cannot but be physical. Now, can the physical sense-data be constituent parts of the physical object they represent? No, they cannot, because had they been so the physical object would have to undergo a diminution of volume or stature due to emission of sense-data from it. If the sense-data are constituent parts of the object in question, then the object cannot have a localized content, i.e. it cannot have a boundary. Moreover, the object will have to accommodate contradictory sense-data in it because the same object may transmit contradictory sense-data to sense-organs of different persons or to sense-organs of the same person on different occasions. Lastly, the sense-organs, though physical, cannot be the object itself because an object perceived to be existing at a distance cannot itself come to a person to affect his sense-organs. For example, a distant tree does not come to the proximity of a person to affect his sense-organs and be perceived.

Sense-data, according to us, are those entities which come in touch with the sense-organs. As the sense-data are not constituent parts of the object perceived, they can only be configurations in the atmosphere that can move on to the sense-organs. An object reacts on the atmosphere adjacent to it and the atmosphere gets configured. This configuration moves on in a chain reaction till it reaches a sense-organ which is affected thereby. In the case of visual sensation, the reaction of the physical object reaches the eyes of an individual in the form of light waves through ether. In the case of auditory sensations, the reaction of the object reaches the ears in the form of waves through air. In the cases of olfactory sensations and the sensations of taste and touch, the objects themselves come in touch with the sense-organs of an individual. The sense-data are thus either physical configurations in air or the physical objects themselves. From the standpoint of an individual, the sense-data are the entities that affect his sense-organs. The nature of effects of the sense-data depends on the physical object, on the intervening medium and on the condition of the sense-organs. The mind which apprehends the effects on the sense-organs analyses and synthesizes them in order to construct the objects. The process of apprehension of an object thus also depends on the conditions of the mind. In view of the above, it is quite possible for an object to transmit to the minds of individuals sense-data of contradictory character though it itself does not thereby become an object of contradictory character. The sense-data may appear to be contradictory but for this the object is not required to be a cluster of contradictory sense-data.

The consciousness of an individual is concerned with the mind and its changes only. The physical objects, the sense-data representing them and the effects of sense-data on the sense-organs lie outside the mind but the objects constructed by the mind are within it. The mind comes to know the existence of these three sets of entities by inference. In order to ascertain whether its inferences and constructions are correct, the mind can apply two criteria, the criteria of workability and of coherence. If these two tests show that the inference and construction of a physical object is not erroneous then the sense-data and their effects on the sense-organs do correspond to the physical object outside the mind.

As the mind itself cannot produce the sense-data at will, the sense-data must have come to the mind from outside. Whatever may be the real nature of the object, the sense-data have come from, to the mind, the object is only of the nature the mind finds in its constructed object because the constructed object works as calculated and expected. For the mind, the object, as constructed by it, is real. Whatever is known is known through the mind. Whatever can be known can also be known through the mind only. A thing cannot be known unless it is reflected in the mind. When the mind knows a thing, it knows not only the thing but also the thing as something lying outside the mind. The changes that the mind undergoes mean that the causes for such changes lie outside the mind. If, on the other hand, the causes for changes lie within the mind, then the mind will be able to make changes in it and create objects at its own will. As it is not so, the causes for changes in it must be external to and independent of it.

The mind analyses and synthesises the sense-data, constructs the physical object and conceives it as existing outside the mind. If this object works as per expectation and if it coheres with other such objects, previously accepted as real, then the newly constructed object can be treated as real. Whenever the constructed object does not respond as per expectation, the construction is modified by the mind. There is no reason for the mind to believe that the constructed object is only a fiction. If the said object works in the manner as expected, then for the mind it cannot but be real. The object is real but real as a construction of the mind or rather as a reconstruction of the object with the sense-data produced by the object earlier.

The mind is concerned only with itself, the changes effected in it by the sense-data, and its construction of objects. The physical object constructed by the mind is within its fold but the said object refers to its existence 'outside' the mind. The construction itself is incomplete without the element of its notion of being 'outside the mind'. Though the notion of the object is inside the mind, the mind is also conscious that the object that the notion refers to exists somewhere outside the mind. The mind of an individual is concerned with the changes effected in it by sense-data and the objects constructed by it but the body and sense-organs of an individual are concerned with the mind on the one hand and with the objects on the other. To the mind, of course, the body of the individual is also a constructed object only.

To an individual there are two worlds: the world of its own mind and the world of physical objects outside the mind. The physical objects are outside and independent of the mind and the mind can have notions of them only through its own constructions out of the sense-data and impressions on it. The mind cannot ascertain conclusively whether its constructions of physical objects do actually correspond exactly with those objects. But for all practical purposes such constructions of the mind will be real if they work as per expectations. Coherence as a test of truth follows next. But coherence of new constructions with previous constructions can be ascertained only after an application of the criterion of workability. In order to ascertain whether an object constructed by the mind is coherent with other objects, the mind is to determine whether the working of the newly constructed object is in harmony with the workings of other objects. Thus, coherence between objects is ultimately determined by the test of workability. Similarly, the test of correspondence is also not an independent criterion. Correspondence between objects constructed by the mind with the actual objects outside the mind can be determined by the test of workability only. If an object constructed by the mind works as expected from it, then it corresponds to the object outside the mind. It is so, at least for all purposes of the mind. To a mind, its constructed object is the only object it deals with. As regards real correspondence between an object constructed by the mind and the object outside the mind, a mind in its ordinary state of existence has no means to be sure of. The mind can ascertain such correspondence only when the object does no longer remain outside the mind. It is possible only when the mind becomes one with the object in question, i.e. when the mind enters into the inner being of the object and apprehends its inner nature. This is the intuitive way of apprehension and not the logical. The logical way of apprehension does not lead to a conclusive ascertainment of truth.

Now, if the body and physical objects do not exist outside the mind can then the mind construct them? According to us, it cannot. It cannot, because, if the body and physical objects are only mental constructions with no objects outside to correspond to, if the mind is a self-contained entity having no principle of communication with the "outside" then it cannot also have any notion that there is actually an "outside" of it. A mind can have the notion that there is an "outside" of it only if there is really an "outside" with which the mind has some communication. Had the mind been a self-contained entity it could never have any notion of "outside" because its notion of the body and physical objects includes the notion of their existence outside the mind. If it is a fact that whatever happens in the mind has also their origin within the mind, then the mind will be a self-contained entity and as such it will not have any notion of "outside". As the mind has a notion of "outside", the impressions in it must have their origin outside the mind. Secondly, if the body and physical objects are only constructions of the mind, within it, and dependent on it, then the mind could be able to produce them at will. In that case there would have been no wants, no dissatisfactions, no pains and no miseries in the life of an individual. The individual could then have everything he wanted. He could have shaped his body as he liked and could avoid bodily pains and sufferings. He could have produced other materials for other purposes. As he cannot do so, it must be said that the mind cannot produce things at will. An individual may at best compound materials in different manners according to his will but the materials in question are independent of the mind. These materials must therefore exist outside the mind. Thirdly, like the critical realists we also say that if the body and the physical objects have no existence outside the mind, if they are purely mental constructions, then they cannot act as they actually do. How can a mental entity roll on the street like a wheel, explode like a bomb or fight like a bull? The ideas of wheel, bomb and bull cannot roll, explode and fight. It may be said that rolling of the wheel, explosion of the bomb and fighting of the bull are only ideas like the ideas of the wheel, bomb and bull and as such the ideas do not roll, explode or fight and that they only refer to those incidents. It cannot be denied that the wheel and its rolling, the bomb and its explosion, the bull and its fighting are all ideas in the mind but these ideas themselves refer to existence of the wheel, the bomb and the bull outside the mind. These ideas depend on the mind only in the sense that for a man they cannot exist independently outside his mind. But a mind itself cannot produce these ideas, because in that case a mind could have avoided entertaining the ideas that annoy it. For example, a mind in that case would have avoided being run over or wounded by bomb-splinters or being gored by a charging bull. In imagination a mind can, at will, compound a set of ideas coherently but this set may not work. The materials for construction and compounding of ideas are the impressions in the mind but the mind cannot create these impressions at will. Even if it is said that the impressions are not due to causes outside the mind then also it must at least be admitted that they exist somewhere, beyond the control of one's will. The causes of impressions must in that case be in the sub-conscious or unconscious region of the mind. But in both these cases the causes of impressions will be within the mind and not "outside" it. The mind, according to us, cannot, in that case, have a notion of "outside". But, in fact, the mind has such a notion. Lastly, if the notiòn of "outside" is nothing but an idea only, then also this idea works, is coherent with all other ideas and as such must be considered as an idea that corresponds to a fact. Even if the said fact is no more than an idea only, the mind, in its ordinary course of existence must, for all practical purposes, consider this idea as a factual entity. Even if the causes of impressions in the mind are not "outside" the mind we must consider them to be so because the notion of "outside" serves all practical purposes and coheres with other ideas corresponding to objects "inside" and "outside" the mind. In the case of illusions etc. the objects constructed by the mind do not work and as such are unreal. Even if the notion of "outside" has no counterpart outside the mind, the said notion works and coheres with all other notions of the mind. The notion of "outside" can therefore be considered for all practical purposes as real.

Knowledge is possible for a mind only, because in every knowledge situation mind must be present as a knower. In a knowledge situation there must also be the object of knowledge, or the known. A knower's apprehension of the known is the knowledge for him. Knowledge thus always refers to a person. Knowledge is knowledge for a particular mind only. But as living beings of a particular species have their bodies and minds similar in structure, the impressions on their bodies and minds, affected by sense-data, and their constructions of physical objects are similar in type. The apprehensions and ideas of physical objects being similar, the experience of one individual can be communicated to another individual under the same species. But though for all practical purposes knowledge is communicable to each other under the same species, the impressions of the physical world on different sense-organs being different, the said knowledge cannot be universal in the strict sense of the term. The impressions on our mind and construction of physical objects may be similar, they cannot be the same or identical. Now, can this non-universality of knowledge be also true in the case of scientific and mathematical knowledge? According to us, the scientific and mathematical laws are universal because these laws are only deductions from the laws of thought. The laws of thought are the principles that work behind the functioning of the mind. The principles that work behind the mathematical truths like 2+2 = 4 are derived from the laws of thought. As the laws of thought cannot be refuted by us, the mathematical truths will be universal. If we substitute physical entities of the world for the mathematical symbols, we find that the physical world conforms to mathematical truths. Sometimes, we find out mathematically the causes that work behind events in the physical world. The applicability of mathematics in the physical world shows that the principles of the mind are also the principles of matter. The physical world conforms to mathematical truths because the principles of matter conform to the principles of mind. The laws of mind being the same for everybody the laws of mathematics are the same to everybody, or, in other words, the mathematical truths are universal. The laws of physics and chemistry are laws of matter and as such are ultimately determinable by the mathematical laws. All other physical sciences like geology, astronomy etc. are sciences derived from physics and chemistry and so mathematical principles are applicable in their laws also. But in the case of life sciences mathematical principles are not always applicable. Mathematical principles do not determine events in the sphere of life. It is so, because in the spheres of life the will of the living being also plays its role and determines events. Though in the case of a living being the pattern of its behaviour can be calculated beforehand, though inexactly, determination of events in the mathematical sense of the term is not possible in the said sphere. An individual may not act according to the laws of mathematics if his will prompts him to act otherwise. The individual will, according to us, is a principle of freedom and in the activities of an individual, the said will is a regulating factor. The biological laws are thus only possible patterns of behaviour of the living beings and the possible physical and chemical changes in their bodies due to reactions of foreign substances on them. The laws of physics and chemistry are general truths relating to the impact of energy corpuscles with each other determinable by the laws of mathematics. The said corpuscles have no will that may interfere in their movements and as such their movements are accurately determinable. Though in the activities of living beings the individual will interferes as a determining factor, such interference is limited to some spheres of their activities only. And in the spheres of their activities where individual will is not operative the laws of physics and chemistry are applicable and such activities are determinable by the laws of mathematics. Though the universal scientific laws express principles that govern the outer world of energy, the principles of energy as such are not expressed thereby. Universal principles that work behind the movement of energy do not exhaust all truths about energy. Scientific knowledge of energy is not all that can be known about it.

Scientific laws are general in the sense that these are related to laws of thought true for every mind. But to discover and know these laws is only to know how energy and entities composed of it behave under certain conditions. Such knowledge is only knowledge of certain principles reducible to mathematical laws and are not exhaustive knowledge of the corpuscles of energy. Complete knowledge of an entity arises out of a total apprehension of the being of that entity and such apprehension is not possible by simply knowing the principles that were followed or will be followed by the entity. Exhaustive knowledge of an entity is possible only when we come in direct contact with its being, or, in other words, be one with it. Only by knowing how the Eskimos live one cannot have exhaustive knowledge about them. To acquire exhaustive knowledge about the Eskimos one must live with them and be one with their community. Knowledge, which means exhaustive knowledge, is thus ultimately personal. The laws of physical sciences, being deductions from mathematical principles, are necessary truths and as such acceptable to every mind. These are principles and laws likely to be followed by the physical entities under certain conditions. But unless it is ascertained that the said entities have actually followed those principles, knowledge about them is not complete. Unless an individual himself conceives the laws of science and their actual application on the physical entities, the said laws cannot be items of his knowledge. Knowledge thus is ultimately knowledge of an individual.

Real knowledge is knowledge of necessary truths because truth must be truth always, everywhere and to everybody. Truth in this sense may lie only in the deductions from the laws of thought and mind. Inductive generalizations are only probable truths because some elements of uncertainty are always present in them. Inductive truths are true in the real sense of the term only when they necessarily follow from the truths underlying energy. Deductive truths, based on the laws of thought and applicable in the case of energy or physical entities, are general and acceptable to all. But to add such truths to one's mass of knowledge the individual concerned is to apprehend them. In other words, an individual is to conceive such truths in order to make them constituent parts of his knowledge. But in the case of energy, an individual cannot claim to have known them because the general laws underlying their movements are only the principles that are and will be followed by energy under certain conditions. As, according to us, there is a difference between a quantum of energy in itself and its behaviour towards other such quanta, to know the behaviour of energy under certain conditions is not to know all about energy. Only by going through the travel account of a country one cannot claim to have seen and known the country. Similarly, by studying the principles underlying the behaviour of energy, one cannot claim to have full knowledge of energy as such. To know energy in itself, one must plunge deep into the realm of energy. And it is possible only if an individual becomes one with energy as such, if he can feel energy as such as if it is his own body. This process of making oneself identical with energy is the process of intuition. Intuition is a super-intellectual phenomenon that arises out of an intellectual sympathy for the existence and activities of energy. An individual at this stage not only knows how energy acts but also knows what energy is. An individual can reach this state when he enters into himself and acquires an exhaustive knowledge of himself. This practically is the state when the individual crosses the barrier of his body, when he is no longer required to guide the body any more, when the process of his individuation is complete and he is no longer an individual. When he knows all about himself, the individual becomes conscious of all the laws and principles of his mind. The individual consciousness is now one with infinite consciousness because it is not now an entity narrowed by the limitations of the body. Acquiring infinite knowledge an individual becomes one with the Absolute, which also includes infinite energy in it. Consciousness at this state is no longer an individuated form of it. It is now consciousness as such, i.e. the original state from which it emerged for individuation. Thus, knowing one's own self, an individual knows all. Intuition thus means perfect identity with the Absolute. This is the state where an individual has full and perfect knowledge of everything in the universe.

Chapter 9

Relevance of Religion to our Hypothesis

In order to ascertain whether the religious feelings and activities of men have a relevance to our hypothesis, we propose to define first the said feelings and activities as also the purpose they are supposed to fulfil. Religion is an attitude of respect and reverence towards the Highest, our creator, and the expression of that attitude in some forms of activities. The attitudes of respect and reverence are feelings of a mind. These feelings are generated out of the conviction that an individual is the wilful creation of a being far superior to him. The other conviction involved in these feelings is that the creator will respond to the feelings of respect and reverence and will be pleased to accept the offerings of men. On being pleased the creator will bestow His blessings on the worshipper concerned and will take care of him in the way of redressing his grievances and helping him in his efforts to attain a happy life here and hereafter. If we now analyse these notions involved in religion and religious activities we find the following pre-suppositions accepted therein.

The source of our being, the creator, is a spiritual entity. We have not come into being from out of nothing. There is a cause working behind our existence and that cause is the will of our creator. Our creator is not simply a principle that works by necessity and has created us under any compulsion. Our creator is not like a mathematical principle who brings us out of Himself like the deductions from a formula. Our creator cannot simply be an abstract principle, He must be a concrete reality having consciousness and energy in Him. Our creator must have consciousness in Him, or we could have no such thing in us. Our creator must have energy in Him, or we could not have energy in us in the form of our bodies. Our creator must have had all that we have and in this sense He is what we are, though His whole being is not exhausted therein. If we are persons in a certain sense, then our creator is also a person in that sense though His personality includes many more attributes. If we are persons, then our creator cannot be an impersonal being. If our creator is a huge machine that produces beings like us, then it is meaningless to establish a personal relationship with him. If our creator is no more than an abstract principle, then we cannot develop a personal relationship with Him. Our religious consciousness and activities being mental, our creator must be a spiritual entity because He is supposed to respond to our devotion and prayers. If there is to be any meaning in our dependence and submission to Him, then our creator cannot be less than what we are. If we are men in a certain sense, He must at least be a man in that sense, though He may be much more than that.

Our creator maintains a relationship with us and responds to our prayers. If He has no relationship with us, then our religious yearnings will be futile. We offer our prayers to Him with the firm belief that He will respond to our call. We beautify our churches, temples and mosques with the belief that He will come to appreciate our sense of beauty. We chant His names in melodious tunes so that He may be pleased. We burn incenses at our places of worship and prayer so that the fragrant air may please Him. All these attitudes and activities express clearly our belief that our creator has not created us to be only left uncared for. A relationship between the Almighty and ourselves is thus implied in our religious yearnings.

All our religious activities are directed towards the fulfilment of a purpose, the purpose of attaining the ultimate goal of our life. God has created us to fulfil some purpose and that purpose will be fulfilled through the natural courses of our life. The natural course of our life is the movement towards its ultimate destiny. The ultimate destiny of a living being, according to us, is to be an individuality developed to the maximum possible extent. A fully developed individuality, or personality, includes in it infinite knowledge and infinite energy and is thus no less than our creator Himself. To be one with our creator is therefore the ultimate destiny of a living being. The purposes of our creator can therefore be fulfilled when we reach the ultimate destiny of our life and be one with Him. If before creation we had been one with our creator and if we cannot go anywhere beyond our creator, then to be back to Him is our ultimate goal of life. To be away from the course towards divinity amounts clearly to a revolt against the creator. As an act of this type cannot but displease our creator, one's activities towards the path of divinity must please Him. Now, our religious activities are nothing but approaches towards God and as such these activities please Him. In whatever way we worship our creator or pray to Him sincerely, He cannot but respond to our call. Our religious activities thus presuppose the belief that such activities please our creator. Without such a belief our religious feelings and activities amount only to a play of make-believe and cannot have a place in our scheme of life.

Our belief that we have been created by a superior being also makes us to believe that the said being is our father. We therefore feel entitled to place our grievances before Him and to pray for their removal. Our belief in a relationship with the creator makes us bold enough to establish a personal relationship with Him and so when we report to Him everything that happen in our life, we feel relieved. In doing so, we believe that the creator shall divert His attention towards our problems and shall redress our genuine grievances. We believe that a living being on the path towards God shall draw His attention and earn His blessings. We believe that all the woes, worries and anxieties of our life shall disappear due to divine interference. Being the source and stuff of everything, our creator has the capacity to interfere in the course of all events and He will actually do so to suit the purposes of His genuine devotees.

We cannot be fully independent of our creator though He has given us freedom to exercise our will. We can steer our lives to any course we like but when we become conscious of ourselves as "exclusive entities" and of their ultimate destiny we become aware of the limitations of our freedom in the exercise of our will. We come to know that our limited knowledge of the courses of the world does not enable us to determine the proper course for our life. We therefore seek the help of our Lord in this respect with the belief that our prayers will not go unanswered. We believe that He will steer our very will so that our activities may be in tune with the divine music. We shall also get strength of mind necessary to cross the barriers on our way and shall ultimately reach the goal. Without expecting help from our Lord, we can neither submit to Him nor can offer our prayers to Him.

As ordinary mortals we cannot fully conceive of the Absolute who is infinite. We therefore conceive of the latter from our limited humanistic standpoint. We think of the Absolute as also a person like us, though we know fully well that infinitude is His very essence. He is not a limited and finite being like us but we conceive of His limited form only for the purposes of our going nearer to Him. When we think of our creator as also a person we try to establish a personal relationship with Him. The relationship we choose to establish depends on our individual temperament. We choose the relationship that appeals to us the most for a personal approach towards our creator. We thus conceive Him as our father, or mother, or friend or our lover. The humanized conception of our creator makes us feel easy to make approach towards Him and to unburden before Him our load of miseries. We do also sometimes ask our Lord to fulfil our desires as if it is our right to claim such things from Him. In absence of any personal relationship with the Lord religion loses its charm and turns out to be a matter of routine-duty without warmth and devotion. And religion without devotion is only a meaningless gesture of our mind.

Our religious activities spring from the innermost depth of our being. It is a spontaneous outburst of our inner feelings and convictions. Religious activities reveal our convictions about our creator and His relation with us. Religious faith grows out of our ideas of the universe. These ideas may or may not be true but a person on the religious path has his conviction that what he believes is true. Religion demands one's whole being to be engaged in it. In his religious activities a person submits his whole being to the Highest, his creator and Lord. A person cannot submit himself to a being that he does not believe in.

Now let us see whether or not the religious activities of men can have any justification from the standpoint of our hypothesis on the evolution of this universe. The source and first principle of this universe, according to us, is infinite consciousness coupled with infinite energy and we have called this entity the Absolute. All energy in the manifested world has come out of the Absolute. Consciousness in living beings has also come in the similar way. The objects of nature are either energy only or energy coupled with consciousness and these are the inanimate and animate beings. Being animate objects of Nature, we, human beings, are products of the Absolute and in this sense we are justified to call the Absolute our God and Father.

We have shown that the Absolute manifests a bulk of His energy which, through evolution, takes ultimately the form of matter. This matter, through individuation, attains that state when consciousness evolves in it to complete the process of individuation and an animate being is thus created. The Absolute being the source of every manifested entity, matter and consciousness in animate beings are all derived from it. So, whatever may be the form of matter and consciousness in a living being he is within the fold of the Absolute as His constituent part. A human body is composed of matter and is a form of energy. It is a part of the energy that was manifested by the Absolute. This energy being within the fold of the Absolute, our body is in God, or, in other words, our bodies are integrally related to Him. Similarly, our consciousness also is a part of the consciousness of the Absolute. All that happens in our consciousness happens within the fold of the Absolute. One's mind and its thoughts, feelings etc. are all within the Absolute and are constituent elements of the consciousness of the Absolute. The feelings of personal relationship with God are also entities within the fold of the Absolute. By personal relationship we mean the relationship between two individuals. An individual is an entity having energy and consciousness. In this sense God, like us, is an individual and an individual person is a replica of God. If God is the highest individual then a living being is a lesser individual. A living being at the earliest state of its existence has a vague consciousness about itself and is not self-determined. Then in course of development it goes on acquiring knowledge of itself and of its environment. In course of this process a living being acquires ultimately all knowledge of itself and of the whole universe. If, to be the highest individual is to have all energy of the universe in itself and to have knowledge of everything then a living being which has lesser energy and lesser knowledge is also an individual. Also, if a personal relationship is possible between two individuals at the like level of development, the same relationship is not impossible between an ordinary individual and the highest individual. If a personal relationship is possible between two persons, the said relationship is also possible between a person and a super-person. Moreover, being our creator, the Absolute is like our father. If we can have a personal relationship with our father we can have that relationship with the Absolute also. The only difficulty in conceiving such a relationship is that the Absolute is too great for us to make an easy approach to Him. For ordinary individuals, like us, to have a personal relationship with the greatest individual may seem fantastic, but it is not definitely impossible.

As the Absolute creates us to fulfil some of His own purposes and as we are made in His own image, He cannot but have interest in us. As our creator-father, the Absolute must look after our welfare and as such we can have a personal relationship with Him. Our mortal father is a being like us and so we can establish easily a personal relationship with him and we can also see easily what he does for us. But as the Absolute is too great for us He is beyond our easy grasp. We cannot therefore come to know easily how we are related to Him, what He does for us and in what ways. Some amongst us therefore keep the Absolute beyond our mental horizon and think of ourselves as independent beings outside the sphere of His influence. We keep ourselves contended with the belief that the world moves on in its own ways like a machine and we are supposed to adjust our lives with its workings. But a little reflection shows that we are utterly mistaken in this conception. Infinitude of the Absolute may keep Him beyond our easy grasp but does not dismiss our relationship with Him.

We conceive of a person as a being who has energy and consciousness in him. A person in the earlier stages of his evolution is conscious of his own existence only, the requirements of his own life and of his immediate neighbourhood only. In course of evolution and development he goes on acquiring more knowledge of himself and of his environment and this process ultimately ends in self-knowledge and in knowledge of the whole universe. Starting from the stage of performing instinctive acts when a living being reaches the stage of acting deliberately, we call him an individual person. The said person, in course of living his life, goes on acquiring more and more knowledge of himself and of the universe thus enlarging his consciousness by owning bit by bit the whole universe and becomes omniscient, or one with the Absolute.

Now, by the word 'person' we mean a living being who has a freedom in the exercise of his will. A disembodied being cannot, in this sense, be a person. Also, in this sense, a physical object, devoid of consciousness, cannot be a person. Lastly, a living being who has no freedom of his will also cannot be a person. We have shown earlier that energy and consciousness are correlated entities and they imply each other. An entity must always have energy, in some form or other, and consciousness, either explicit or implicit in form. Thus, according to us, consciousness must be attached to a body, or energy, and there is consciousness even in a so-called non-living object. A person has not only a body, consciousness and freedom in the exercise of his will but has all these three in the explicit form. A stone has only an explicit body but it has no explicit consciousness and freedom of the will. A stone therefore is not a person. Plants and lower animals are also not persons, because they have their bodies and consciousness in the explicit form but have no freedom of the will. Their activities are only mechanical or instinctive. Now, if body, consciousness and a free exercise of the will are the criteria of a person, then the Absolute must be a person, par excellence. Being the very source of energy, the Absolute has definitely a body because a body is nothing but energy in a particular form. The Absolute has also consciousness because He is the source of all consciousness. A difference between consciousness of the Absolute and of other beings is that the former has no mind whereas the latter, particularly the beings having freedom in the exercise of their will, have their minds. It is due to the fact that being omniscient the Absolute has no need to have a mind to analyse any data for apprehension. Everything in the universe is within the fold of the Absolute and so, being both the subject and the object of knowledge, the Absolute is not in need of any organ for knowledge. But in the case of a lesser being it is not so. A lesser being is not the subject and the object of knowledge at the same time because he is not an all-pervading entity. As such, he is in need of sense-organs for collection of data of knowledge and also is in need of a mind to analyse and coordinate the said data for apprehension. Though the Absolute has no mind, he is in another sense all-mind, because, constitutionally, mind is nothing but consciousness. Mind in a lesser being is only a centre where analysis and reflection is done for apprehension. In the Absolute this apprehension is direct. The Absolute needs no analysis and reflection to apprehend anything.

There being nothing to counteract on it, the Absolute is free in the truest sense of the term. He is free because there is nothing to limit Him, nothing to act as a barrier to His freedom. His limitations are all self-imposed and these are His own laws and principles. Thus the Absolute fulfils all the three criteria of being a person. If a person is an individuated self-conscious being with freedom of the will then the Absolute is the most individuated and free self-conscious being and so, in this sense, He is a super person. The Absolute can therefore be conceived of as a person, though He is much more than that. If we are persons, then, being our source and creator, the Absolute must also be a person, because He cannot be less than what we are. In view of the above, personal relationship with the Absolute seems to be quite possible and so the religious sentiments and activities of a person cannot be dismissed as only idiosyncrasies. These activities are not illogical to a person who wants to get himself liberated.

The purpose behind creation, as we have said earlier, is to bring out self-conscious creatures who will have freedom in the exercise of their will. These creatures are supposed to weave out and display patterns that are not foreknown to the Absolute. These unseen patterns are to break the monotony of existence of the Absolute. In view of this conception, it may also be said that these patterns must be transitory in nature. These patterns cannot continue to exist in the same state for ever, for then it will mean the creation of a different type of monotony. This scheme must therefore include in it a process that leads towards a reversal of the whole position. So, creation of living beings necessitates dissolution of them. It may, therefore, be said that the creation of the world is to be followed necessarily by a dissolution of it.

The living beings are brought out by the Absolute to fulfil a purpose of His own. The Absolute must therefore have a keen interest in the affairs of the said beings. It is so, more particularly in view of the fact that the fulfilment of purposes of the Absolute depends on the development and behaviour of the said beings. It may also be believed that being responsible for the very birth of living beings the Absolute cannot and will not disconnect Himself from their affairs. As the Absolute reveals Himself through Nature it may be assumed that for living beings, the natural course of growth and development is right and it is what is intended by the Absolute.

It may also be said that the proper course of activities for a living being should be those activities which lead him towards his self-realisation through a free exercise of his will, enlargement of his consciousness and development of his personality. By self-realisation we mean that state of a person's existence when he knows completely his own being. Self-realisation is possible only when a person becomes one with the Absolute. Does one's progress towards self-realisation then mean a process that stands contrary to the purposes of the Absolute? According to us, it is not contrary to what is intended by the Absolute due to the following reasons. Firstly, self-realisation is not an easily attainable state. It is possible for a very few persons only. Self-realisation of these few persons does not mean the self-realisation of all, frustrating the purposes of the Absolute. The persons who attain the state of self-realisation leave behind them innumerable living beings to weave out and display patterns not foreknown to the Absolute. And secondly, the natural course of movement for a bulk of energy and consciousness, which constitute a living being, is only to return to their source, the Absolute, for, there is nothing other than the Absolute for them to move into.

Being a person, the Absolute must have His feelings like those of us. His feelings must be similar to those of us because we have been created in His image. The principles that work behind our feelings cannot therefore stand contrary to the principles that work behind His feelings. Being our creator the Absolute takes interest in our activities and He has His feelings on that account. In the analogy of our own feelings therefore we may say that the Absolute will be pleased at those activities of a person which conform to His purposes. It may also be said on the same ground that the activities of a person which stand contrary to the purposes of the Absolute will only displease Him. To say that the Absolute is neither pleased nor displeased with our activities amounts to saying that the Absolute takes no interest in our activities at all. We are therefore justified in our belief that a person who lives his life in accordance with the laws of Nature which guide him towards more individuation and development of personality earns special favour of the Absolute. Though a person has freedom of his will he is not supposed to misuse that freedom. A person is supposed to exercise his freedom in choosing the right path of his life for further individuation and development of his personality. The Absolute's favour may come to a person in an unpredictable way. In whatever manner the favour may come, it will help the person concerned to attain the state of his self-realisation. On the same ground of reasoning it may be said that a person who goes against the laws of Nature or moves away from the path to self-realisation will have to face difficulties in some form or other. In other words, a person on the right track will be rewarded and a person on the wrong track punished. The thoughts and expectations behind our religious activities are thus not unjustified. We are quite justified to call the Absolute our God and try to establish personal relationship with Him. We can also pray to Him as His devotees.

Now, a person's freedom of the will is not unlimited. He cannot do everything he wants to do. An individual enjoys a relatively greater freedom in his mental activities than in the activities of his body. Limitations in the field of a person's mental activities are limitations of his faculties and limitations of the particular level of his intellectual development. Limitations in the field of a person's physical activities are limitations of his physical power and limitations due to obstructions of external agencies in the form of activities of other physical objects, including other living beings. We thus divide the activities of a person into two types, mental and physical, and propose to study the extent of their freedom in these two spheres separately.

In the case of mental activities the limitations of a person are due first to limitations of the particular stage of his mental development. Apprehension of entities of the world depends on the richness of apperception a person is capable of. In other words, the stock of knowledge already acquired by a person determines the mental activities of a person as also his capacity to apperceive a freshly acquired piece of knowledge. The apperception of a person who has a richer stock of knowledge will be richer than that of a person with a poorer stock of knowledge. The second limitation is the limitation of faculties of a person. To analyse a mass of new data of knowledge and to relate it to the mass of knowledge already acquired is an intellectual exercise and a person has to be capable of it. Capability of this kind is the so-called power of the intellect and it depends on the attention a mind pays to apprehend an object and the stage of mental life a person has arrived at. Power of the intellect depends also on the molecular structure of brain-cells of a person. Thus on account of the above a person may have more limitations in his mental activities or less. Even in the case of a person of highly developed intellect, there are limitations in his act of compounding the sense-data and of conceiving the same in an integrated form. A person can easily conceive of things he comes across in his day-to-day life. But he faces difficulties in conceiving uncommon things. A person has also limitations in conceiving macroscopic and microscopic entities. We, human beings, cannot conceive of objects too great or too small for us. There are limitations also in the act of conceiving abstract entities. The entities we get through abstractions cannot be conceived by a person easily. There are also abstract entities which can never be conceived by a man. The mental activities of a man thus have a good number of limitations. A man enjoys freedom only in a limited sphere of his mental activities. There are elements in the mental life of a man which are controlled by circumstantial forces and also by the Absolute. Being the agent of his mental activities a person has definitely a control over those activities. A person's circumstances also mould his behaviour and as such control his mental activities to a certain extent. Lastly, the Absolute, being the source and sustainer of everything, also controls the mental activities of a person.

The freedom of a man in his physical activities is also not unlimited. A man does not enjoy freedom in all of his physical movements. Some of the limbs and organs, particularly those which are not placed at the peripheral regions of the body, are not within the control of a person. For example, the organs for digestion, blood circulation etc. are not controlled by the person himself. These organs go on working in their own ways and the will of a person is not operative in their cases. Only some organs of the peripheral regions of a person's body are controlled by him and that also with certain limitations. The hands, feet etc. of a person cannot be in movement for ever and these organs cannot also move in any manner a person wants them to move. The period of time these organs can be in movement depends on the energy supplied to them and we know that the quantity of energy in an organic body is limited. There are also limitations in the movements of peripheral limbs and organs due to resistance from other physical objects.

In view of the above we may classify the mental and physical activities of a person into three types—activities that depend on the free will of a person, activities which are determined by his mental and bodily circumstances as also by external agencies, and activities performed by his mind and body on their own accord. The first type of activities depend on the mind of the person, the second type on entities external to his mind (which includes the body of the person concerned) and the third type directly on the Absolute, or Nature. In another sense, both the first and the second types of activities depend on the person concerned because the first type depends on the person's will and the second type is determined by the deserts of the person concerned. Activities of a person in the third type are determined by the laws of Nature or by the Absolute.

In that sphere where a person is free, the will of the person determines his activities. The person in this sphere may choose to follow the course of activities which lead him towards virtuous or towards vicious ends. The path towards virtue is that course of activities which leads a man towards fulfilment of his ultimate goal of life, or his union with the Absolute. The path of vice drags a man away from the course towards his union with the Absolute. Acts performed after deliberation in the conscious region of a person's mind may be called his conscious activities. Unconscious activities are not performed after conscious deliberation.

Of the unconscious activities, mental and physical, not all are instinctive, or activities of the vital principles controlled by the Absolute. Some of these activities are results of unconscious deliberations and as such responsibilities for these activities devolve on the person concerned. These activities may also be either virtuous or vicious. Virtuous activities of a person conform to the purposes of the Absolute. A virtuous man therefore earns the grace and blessings of God. A man on the vicious track earns only the wrath of God and deserves punishment.

When a person is able to deliberate on matters methodically he is called to have developed his rational faculty.

At a particular stage of the development of his rational faculty a man becomes conscious of his own self and its ultimate destiny. Before attaining this stage also a man goes on performing rational acts but only mechanically. A man at this stage is not conscious that he is a person who acts or that he has a destiny at all. But when a man becomes conscious of his separate entity as a personality and when he begins to think about the value and significance of his life, he is called a self-conscious person. According to us, this consciousness of one's 'self' is the root of religious consciousness in a man. When a man becomes conscious that he is a person with unique identity he tries to determine his position in his immediate neighbourhood and in the universe. He becomes conscious of his own individuality and begins to think always in terms of his own self and its unique position in the universe. He tries to determine what is good for him and what is bad. These thoughts, though basically selfish in nature, lead him towards the knowledge of phenomena (Science) and noumena (metaphysics) and terminate ultimately in art and religion. We discard the view that fear is the origin of religion. In support of our view we like to point out the fact that the emotion of fear is present not only in men but also in the lower animals in a more acute form but these animals have no religion, in our sense of the term. Fear can at best incite a person to avoid the source of fear. Fear dissociates a person from the source of fear and not associates him with it. But religion, we know, is the association of a person with the supreme power, or God. Religion, according to us, is based on a feeling of love for the supreme power.

When a man becomes conscious of his individuality, his ego, the "I" in him, he feels the difficulty of conceiving 'his absence in the past or in the future.' He cannot conceive that once he "was not" and that once he "will not be". A man cannot think of his absence without the presence of a principle of consciousness that knows that absence. Thus logically, a man cannot conceive of his own absence. A man therefore goes to think that he was, he is and he will be. This idea of man's existence before birth and his existence after death conforms to the metaphysical truth expounded in our hypothesis on the evolution of this universe. A man, according to us, is a principle of consciousness evolved in a body, or in a cluster of energy. The principles of consciousness and energy in a man are only the parts of universal consciousness and its implied energy. Being the source and first principle of the universe, universal consciousness and its implied energy never get annihilated and as such individual consciousness, being a constituent part of the first principle never gets annihilated. A man's self in this sense is thus immortal. It had no beginning and so will have no end. A man's idea of his ever-continued existence does therefore stand as a fact.

In course of his metaphysical quest a man comes to know the origin and destiny of his individual self. He comes to know that his individuality has come into being through the will of the Absolute, who is his creator and God. As the Absolute is the source and first principle of this universe, the ultimate destiny of a living being cannot but be its union with the Absolute. The ultimate destiny of all entities in the universe is to return back to the Absolute, who is a principle of unity in diversity. As the Absolute is a principle of omniscience, omnipotence, omnilucence etc., to be one with the Absolute is to acquire these qualities. All efforts in a person's life should therefore be directed towards development of these divine qualities.

A man's consciousness is in essence the same as universal consciousness but for all practical purposes it is only a veiled form of the latter. This veil is the veil of the limitations of an individual. Consciousness evolves in a cluster of energy to guide the latter in its movements towards individuation. As without a calculated course of action the individuality of the said cluster of energy may get lost, a principle of intelligence is necessary. Consciousness provides that intelligence and engages itself in the services of the cluster of energy, or the body of the individual. Though consciousness, as evolved in a body, is an entity identical with the consciousness of the Absolute, it limits itself by confining its activities to the fulfilment of purposes of the body only. Being only a part of infinite consciousness, individual consciousness reaches its destiny only when it overcomes the limitations of its individuality and attains infinitude again. To cross the barrier of its narrow individuality an individual is to go on acquiring knowledge till it reaches the state of infinitude again. In this state of a person's existence, his narrow individuality will get lost in the greatest individual, the Absolute.

Consciousness evolves in a cluster of energy to maintain the principle of individuality in it. The very basis of individual consciousness is thus narrow and selfish. This narrowness in outlook and selfishness in activities are the reasons for stagnancy in the progress of individual consciousness. These are also the roots of all evils in human life. Selfishness gives rise to perverted self-love and false love of power. Selfishness is the basic nature of a person, because, as an individual, he is confined to his individuality. An individual may continue to make efforts to enlarge his individuality but till the process of such enlargement is complete, he remains an individual. And to be confined in one's individuality is to be selfish, after all. But a man may perform his selfish acts in two different ways. He may act in such a manner that his activities do not hinder the process of enlargement of other individualities. He does not in that case come in conflict with other individuals. An individual may also act in ways that hinder the process of enlargement of other individualities. The first type of activities, though basically selfish in nature, helps in the normal development of a man's mind and develops the individuality of the man. But activities of the second type engage a man in conflicts with other individuals and destroy the tranquility of his mind. These activities distract and diffuse a man's concentration of mind, giving rise to the consequential evils which hinder the process of enlargement of a man's individuality. Real knowledge that enlarges a man's individuality cannot be acquired by a mind which is not in a pacified state. Selfish acts of the second type give rise to perverted self-love and a lust for power. Perverted self-love impels a man to deprive other persons of their basic rights as social beings. Perverted self-love makes a man run after wealth even at the cost of sacrificing human values. A man under its spell loses the integrity of his character and balance of his mind. He loses all control over his mind and cannot therefore make any progress in his spiritual life. Perverted self-love develops in him a lust for power. He wants to enjoy the pleasures of dominating over others. To fulfil this desire he goes on violating the basic laws of social life, disregards the human values and creates social evils. Perverted self-love instigates a man to run after bodily pleasures, like those relating to sex etc. even by destroying the ethos of a society. Uncontrolled sex-urges unhinge a man from his self-composure. It destroys peace and tranquility of a man's mind and brings him down to the level of beasts.

Some evils are created by men through their activities under pressure of circumstances. These are evils created by a man in order to avoid creating greater evils. But whatever may be the reasons behind evils created by a man, he will be responsible for them as, after all, his acts have created those evils. After all, it is his inability to sense danger in the circumstances he plunged into. His lack of knowledge being the root of such evils the person concerned cannot shirk off his responsibility in this regard.

There are some more evils that are created by a man through his acts performed in sheer ignorance of the possible consequences. Ignorance cannot make an individual free from the responsibility involved in creating evils. Moreover, all evils, created by a person are, in the last analysis, due only to his ignorance. Even in cases where a person creates evils knowingly, he does so being ignorant of the fact that such acts are detrimental to the purposes of his own life.

Evils, created by a person, obstruct the natural course of events, particularly the growth of life and individuality, aimed at by the Absolute. To go against the Absolute or Nature is to be involved in an anti-life activity and as such it obstructs the growth of a person's individuality. Evils create conditions that destroy the peace and tranquility of an individual's mind. Evils beget evils. A person who creates evils invites the fear of being retaliated.

Evils created by a man fall back upon him. Evils are the results of abnormal activities of living beings and as such they bring in sufferings for them. Abnormal activities can bring in only abnormal episodes in the life of individuals. A person behaves abnormally because he does not know that the purposes of his life are fulfilled only when his activities are in tune with the activities of Nature. Men suffer because they do not know how to live.

A man who suffers begins to think about himself. Sufferings make an extrovert mind introvert. A man who suffers begins to think about his own faults, miscalculations and inabilities. He tries to rectify the defects in his own activities so as to avoid future sufferings and in so doing comes to know that his powers and abilities as an individual are very much limited. In course of analysing himself and his endowments, a man ultimately comes to the question of his origin. He starts conceiving the principles of causality and the principles of creation and evolution of the world. In course of this process of thought a man becomes conscious of the supreme power that presides over the universe and is the source and stuff of everything. Being the creator and sustainer of everything in the universe, this supreme power is our God-father. All these mental activities of a person thus develop a religious consciousness in him. This consciousness impels the man to be engaged in religious activities for his ultimate union with the Absolute.

Sufferings not only make a man conscious about his own self and about God but also make him conscious about the fact that God, being his creator, cannot remain indifferent towards him even when he approaches Him in the right spirit and in the proper way. If he is in the right path and if he does not create obstructions in the ways of God then he will surely earn His grace and blessings and will be made by Him free from his sorrows and sufferings. If the relationship between God and man is a relationship between the creator and the created then God must keep His connection with men. A man therefore tries to approach God either as His son, or as His friend or lover or as His servant, with the belief that his sincere call must reach His ears and He must redress his grievances up to the extent he deserves.

A man approaches God not only for the purpose of becoming free from the sorrows and sufferings of life but also for the purpose of enjoying a smooth journey towards his ultimate destiny. A self-conscious person comes to know fully well that it is very difficult for him to choose the right path for his life's journey. Even when he is on the right path, it is very difficult for him to avoid the harmful allurements that detract him from the said path. A self-conscious person therefore seeks the guidance of God in his life's journey and submits to Him in right earnest. He knows that when God Himself steers the course of his life he will commit no mistakes and will get strength necessary to face the hard struggles of life.

A person can approach God only through the instrumentality of his mind, because, his very apprehension of the idea of God is done through the mind. And the easiest of such approach is the approach through prayers in a true devotional spirit. To develop the spirit of devotion a person is to keep his mind under control. He must not be swayed by the bodily passions and must keep the body in its normal order so that it may not disturb the mind unnecessarily. A prayer in sincere devotion connects the devotee to God. If the creator is related to the created and if the created being cannot approach the creator physically, then the only way left to the created being is to approach the creator mentally, through submission and prayer. To be a sincere devotee of God and to draw His attention and response one must apprehend divinity intellectually and must employ himself in the services of God. A man is very much engaged in the affairs of the external world. To withdraw his mind from the external world, and to direct it towards God are therefore not easy tasks for him. Moreover, to be in unison with the ways of God, a true devotee is supposed to abstain always from creating evils. Evils are not created by God because there is no justification for Him to burden His creatures unnecessarily with sorrows and sufferings. Just after being created a creature is not required to suffer because he has not yet performed any act to deserve punishment through sufferings. If God cannot be made responsible for the creation of evils, then the evils can only be the results of the misdeeds of created beings. Now, in order to deserve the blessings and loving care of God, a devotee is supposed to abide by His will. To get responses to his prayers, therefore, a true devotee is required to live a pious life by engaging himself in the services of God.

A genuine prayer, on analysis, reveals its following elements. A genuine prayer can come only from the heart of a man who has a firm belief in the existence of God. A hypocrite cannot go near to God. Mechanical chanting of words is only the futile gestures of a religious sceptic. These gestures do never bring a person near to God. A sincere prayer to God is a total submission to Him. Unless a man submits his whole being to God he cannot be sincere in his prayers. Prayers seeking help from God implies a shifting of the burden of one's life's toils to God and it means a submission of one's whole being to Him, without reservations. A submission of this type includes also the surrender of one's freedom of the will. God must have the whole being of a devotee at His disposal so that He may mould his very being and remake his fate. A true devotee makes his will the will of God.

A genuine prayer is the outcome of a devotee's spirit of renunciation of transitory affairs of the world. A man unduly entangled in the affairs of the world can neither divert his attention to God nor can surrender to Him. One can pray to God in right spirit in silence only. To be a devotee does not, of course, mean that one is to be a mendicant. God does not create man with the intention that they all would renounce His world. The world is not to be renounced by men, its value is only to be assessed by them correctly and utilised as means for their ultimate union with God. God has given us freedom of our will not with the intention that we shall misuse that freedom and go away from Him. Our worldly life is not an end in itself and as such we are supposed to go beyond it. We are to utilise the world as a means towards our ultimate end and not to spoil our life for it.

Another element in a genuine prayer is a real love for God, a real aspiration and craving for union with God. Without this element a prayer can only be the gesture of a hypocrite. A man cannot seek for an union with God unless he loves God. If a man is indifferent towards God or is afraid of Him he cannot have the desire to go near to Him or ask for his help. A man must love God as he loves his kith and kin, or, he cannot depend on Him. Being our source and creator God knows only too well when we pray to Him sincerely and when not. God can in no case be duped with one's hypocritical gestures. He only helps them who deserve His blessings.

The ways of God in extending His helping hand towards a devotee may not be discernible to him, because God may not act in the ways we act. The devotee may not therefore be aware that he has won the favour of God. He comes to know that God has responded to his prayers only when he observes that some impossible events have somehow occurred in his life. If an event was likely to occur in his life and it actually occurred the devotee may not be aware that he is under the loving care of God. But when an event occurs which was not likely to occur in his life the devotee comes to know that God has responded to his prayers. Being not directly aware of the presence of God, a devotee does not know the ways God acts in and as such God's ways appear to be strange to him. From the standpoint of our human ways of looking at things the devotee observes that an impossible feat has been made possible by God and so he calls such acts of God, miracles. But these so-called miracles are not actually miraculous in the real sense of the term. These events may appear to be strange to us but are not surely impossible, or, how could God make them possible? These acts of God seem to be impossible to us because we can neither perform such acts nor can conceive of them. The limitations of our faculties confine our activities to a limited sphere and so we use the term "possible" for those activities only which are within that sphere. Activities which lie beyond the said sphere seem to us to be impossible and so we call them miracles. God makes only the possible events actual, though these events may seem to be impossible and strange to us. The word "miraculous" is thus only a misnomer. A miraculous act means only a supernormal act. A supernormal event in the life of a devotee makes him conscious that the supreme power has interfered in the affairs of his life. This apprehension creates in the mind of the devotee a feeling of elation and makes him keener towards his advancement in the path of devotion. It also creates a feeling of comfort for the devotee in his distressed condition.

Being the source and stuff of everything in the universe, the Absolute or God is immanent in the world. The world is in God and God is here in the world. The world we live in is a partial manifestation of God. The material and psychical aspects of it are nothing but God transformed into these aspects of His being. But as God's total being is not exhausted in the world we live in, He is not present here with all His splendour. God is present in the world but the world is not all divine, because the world contains elements contributed to it by living beings also who are yet to be divine. Though our world has all the potentialities of becoming divine, it is yet to be so. God's being, according to our hypothesis, is infinite consciousness coupled with infinite energy. Our conception of becoming divine, therefore, means 'to have energy with self-consciousness, evolved in it'. This self-consciousness, we have said earlier, leads to the attainment of infinite consciousness. Physical entities, which imply the presence of consciousness in them, are yet to evolve that consciousness and as such are yet to be divine. Animate beings on the lower scale of evolution have consciousness but in an undeveloped form. A long way lies ahead for them to be divine. In the scale of evolution and development the human beings are at an advanced level because they are self-conscious to a certain extent, but most of them utilise their energy and efforts in the affairs of their bodies only. They have no urge to cross the limitations of their outlook and to make progress in the path towards divinity. They do not explore the potentialities of their knowledge and power and as such do only remain in a narrow sphere. Every man has the spark of divinity in him but very few of them turn this spark into a fire to have its warmth and light. Being indifferent towards his ultimate destiny, a man lives his life in the wrong way and thus drives himself away and away from the path of divinity. Real wisdom, real power, real freedom and real joy remain outside the sphere of such a man's life. Divinity and its nectar remain beyond the reach of such a person and God is therefore transcendent to him. But to a person who has crossed the limitations of his body, mind and environment, God is not so. In course of enlarging his personality such a man goes on realising divinity and moves nearer and nearer to God. He is in actual communion with God and he realises that God is present everywhere. To him the world appears to be a manifestation of God and not simply a dance of atoms and molecules or their epiphenomena.

In the light of our discussion in this chapter we can, perhaps, now suggest that the basic tenets of religion and religious ideals do not in any sense run contrary to our hypothesis on the evolution of the world and man's destiny in it. Religious activities of men are nothing but philosophical truths adopted and practised in life. Leaving aside the minor details of religious practices, the ideals of religion cannot be called crazy ideas of idle brains. From our standpoint, the religious activities of a man are not idiosyncrasies but the only possible means to attain the *summum bonum* of his life.

Index

A

Abnormalities, 135-36

Abode of living being

evolution, 180

Absolute (the), 163, 226-29

laws schemes formulated by, 72, 74, 79, 81, 87

manifestations of, 168-69

levels of existence, 173

manifesting feeling of want, 142

manifestation of energy, 143

aim of, 143

knowing/controlling/power, 147

modification of, 141 infinite energy of, 141

will of, 104

manifestation of, 105-106

principle of will, 106

energy manifestation, 225

purpose for living beings, 230

Abstract entity, 62, 233

Accidental action, 62

Action, 62

governance of, 62

Activities

enlarging individuality, 237-38

types, 234

Activity (4), 164

Affection

in sense organs, 199-200, 207-208

and ideas formed in mind differences, 199

in the mind, 208

Amoeba

sense organs in, 151

Analyzing, 165

Animate entities movement, 145

Appreciation of beauty, 153-54

Apprehension, 205

of activities, 232

Apprehensive principle, 195-96

concerning body condition, 197

Artist (being an), 163-64

Artistic creations, 154-56, 158-59

ontological status, 157

basis of, 157

coherent form, 158-59

Artistic sense in creatures, 152

Assimilation, 165

Assuming of body form, 137

B

Beauty

and harmony, 152-54

appreciation of, 153

Behaviour of energy, 75
Bergson, Henri, 11-13, 19, 24, 35
Berkley, 21
Biological laws, 217
Bodily affairs/desires, 203-204
Body
 growth and adjustment, 132
 functions, 127-28
 of an organism, 100
 relation with mind, 204

C

Calcium
 absorption in body, 119
Carbon compounds, 112
 individuated of, 112
Carnivores
 evolution, 79
Change, 71, 73-74, 175
 principles of, 110
 meaning, 168
 as energy principle, 171
 and transformation, 78
 pattern and individual, 148
 implications, 176
 in entities, 176
 undergone by mind, 211
Chemical elements, 64
Chemistry laws, 216
Circulatory system, 127
Coherence, 212
 between objects, 213
Collection/protection of food, 94
Competition and struggle, 183
Compounding of ideas, 214-15
Conceiving things/ objects, 232-33
Conflict and struggle, 183, 185-87
Conscious life, 13
Consciousness, 13-14, 23, 62, 90-92, 101, 106, 112, 114, 119, 165, 167, 236
 and matter, 193
 in forming sense organs, 196
 manifestation, 193-94
 regions, 195
 and energy, 194
 and mind, 200-202
 and instincts, 200
 basis, 237
 evolution, 237
 finite to infinite, 207
 in organic compounds, 161
 of a living being, 35, 37, 39
 hypothesis, 36
 power of choice, 38
 supra/prime, 39-40
 life, 38
 of absolute difference with consciousness of organism, 117
 of self, 235
 presence of mind in, 194
 division of, 115 limitations of, 103
 meaning of, 168
 levels, 177
 and change, 171
 discrimination by, 119
 states of, 11
 and life, 92
Contemplative activities of mind, 101-102

Continuity of Absolute, 168
Conversion of materials, 126
Corpuscle
meaning, 108
Corpuscles of Energy, 64, 73, 81, 91
laws governing, 65
principles of, 79
equilibrium, 87
Corpuscles of energy
manifestation, 174
changes in, 178
forces of, 109
movement results, 193
movements in, 172, 178
and space, 178
and principle of differentiation, 174
forces working on, 174
Course of action
suitability, 61
Courses of nature
following, 187, 189
Courses for undergoing change, 52
Creation
purpose behind, 229
Creative activities
of living beings, 148-50

D

Deeper aspect of mind, 21-22
Deformities, 135
Delebration, 61
of the mind, 99
Destruction of animals181
Detention of feelings, 39
Development of energy, 145
Devotee (genuine), 241-42
Death, 103
and consciousness, 103
Differentiating entities, 98
Differentiation
and integration, 144
manifested energy, 82
of a chemical compound, 179
Digestive system, 127
Direct apprehension, 205
superiority, 206
Discrimination
increase in, 116
work of, 113
Discrimination/integration, 130
Discriminative activities, 96, 99
Divinity, 244
Duration within us, 12
aspects, 13-15

E

Economic activities, 161-62
Ecstatic feelings of poet/musician, 29
Ego-I-Consciousness, 235
Ego-sense, 145-46
Electrons, protons
movement, 83
equilibrium, 83
Element of probability, 86
Elements
in movement, 26
necessary for maintenance of life, 94

Emergence of life, 79, 110
Emotions
aspects, 13
Energy 59, 62-63, 65
integration of, 115
meaning, 168
manifestation, 169-70, 172-73
elements, 173
expenditure, 170
in human body, 226
knowledge of, 219
corpuscles, 64
and consciousness, 228
implicit, 170
principles(3), 171
spending, 60-61
in prime consciousness, 60-61
manifestation, 61
division, 65
evolution, 65
Enlargement of individuality, 237
activity of, 237
Entities, 55
outside our mind, 24
implicit/explicit, 53
in consciousness, 195
forms of, 195
knowledge of, 218
(being an), 164
manifestation stage, 143-44
Environment
creating, 182
Environment/conditions
of living creatures, 148, 150
Equilibrium state, 87
Ethics of utility, 189, 191
codes of, 191
Evils created by man, 238
Evolution, 50, 105
and the process of formation, 88
and laws of change, 88
principle of, 105
of consciousness of energy, 120
of energy to carbon compounds stage, 117
four principles of, 117
of life, 90, 98, 110, 113, 123
principles guiding, 92
of mind purpose, 203
of organism, 80
of Space, 54, 56
Principle of, 59
Exclusiveness and privacy, 107
Exercise of a physical sense organ, 27
Exist(to), 59
Existence
and non-existence, 59
and work, 60
Experiences, 21
Explicit
and implicit form of consciousness, 43-44, 47
energy, 63, 65, 69
entity, 54
form of time consciousness, 47
Exploring world, 185

Extension, 20
External world, 25, 27
Extra-mental objects, 26
Extrovert nature
  development of, 185

F

Faculty of exercising a sense organ, 27
Feelings
  and consciousness, 17
  state of, 30-31
  agreeable/disagreeable, 31
  and person, 231
Fight to exist, 181
Food intake/assimilation, 131-34
Forces working on corpuscles of energy, 71
Formation, dissolution, reformation, 88
Four forces of movement, 145
Four principles of evolution of energy, 117
Four principles-operation of, 146
Freedom of the Will, 111, 163, 231
Function of body, 123
Fundamental self, 12

G

Genuine prayer (a)
  features/elements, 241-42
God
  and man relationship, 240
  approaching
    reasons for, 240-41
  consciousness, 240
Good and bad acts, 188
Growth of living creatures
  controlling of, 149
Growth of systems in body, 134

H

Herbivores
  evolution, 79
Homegenius
  time, 16, 20
  duration, 13, 15
  space, 19
Homogeneous time, 20
Human body as form of energy, 226

I

Idea, 24, 68, 197
  of movement and time, 25
Ideas
  aspects, 13
  of the mind, 214-15
  as symbols of entities, 197
  of matter, 24
Images constituting matter, 22
Implicit and time consciousness, 43, 45-46
Implicit
  energy, 170
  entities in prime consciousness, 59
  to explicit entity, 53-54
Impressions in the mind, 215-16
Impulse of pure activity, 23
Inanimate entities
  movement of, 145
Individual
  and world change power, 147-4

freedom, 148
consciousness, 51

I

Individualities
development of, 184-85
enlarging, 237
consciousness, 235
Individuals
free will, 147-48
Individuated activities
and Absolute, 108
Individuation, 90, 92-93, 96, 130-31, 161-62
development, 201
of energy, 137
and appearance of life, 130
principle, 132, 172
of entity, 145
Individuauted energy
need of consciousness, 194
Infinite
attachment, 201
consciousness/energy, 69, 167, 202
knowledge, 219
acquiring, 219
Inhalation of air, 134
Inner/outer regions of mind, 125
Inorganic to organic
compounds
transformation, 86-87, 89, 111, 190
connect, 81
Instinctive activities, 38, 98
Instincts, 182, 186
and urges, 136
working for maintenance of life, 94
Intellect, 21
power of, 232
Intellectual oneness with entity, 202
Intellectuality, 37
Intelligence inclinations, 17
Intelligent beings, 188
Intense feelings
in love/melan-choly, 29
Internal/outer organs role, 124
Intuition, 21
Intuitive knowledge/apprehension, 202

J

Journey of mind, 29

K

Knowledge of oneself, 203-204
Knowledge of self/universe
acquiring, 186
Knowledge
of the environment, 204-205
of body, 204
Knowledge situation, 215

L

Law
implication, 76
Laws
of energy/science, 81, 202
specific for energy, 84
Leaders of species, 183
Life
manifestation of, 85, 123
principles, 86-87

Life-mind complex, 123, 127
two fold functions, 129
outer-inner aspects, 125
function of, 118-19
and teeth growing in body, 120
Limbs evolution of, 180
Limitations of body, 137
Living and non-living compound
differences, 129
Living being
struggles, 179
phases, 179-80
necessities, 179
evolution conditions, 149-50
purpose of, 230
Love/melancholy
feelings, 30

M

Manifestation
as a principle, 172
meaning of, 172
Manifestation of energy, 72-75, 81, 84, 108, 172
and consciousness levels, 177
Material needs
minimizing, 206
Materiality, 19
Mathematical faculties, 27
Matter
perception of, 19
views on, 21-23
Maturing of mind, 150
Means for sustenance, 181
Mechanical forces guiding nature, 80
Mechanical laws, 86
Medium Conceiving
between prime consciousness and energy, 65
Melodies, 152
Memories, 11-12
Mental
activities, 233
freedom of, 232
Mental entities, 24
Mental life working, 19
Mental/physical activities
types, 234
Metaphysics
and religion
entering domain of, 185
Mind, 97
phases in progress, 103
contemplative activities, 101-102
disciplining of, 102
aspects of, 21-24
observation by, 150
study of Universe by, 150-51
Mind (contd.)
use of sense organs by, 151
functioning, 151
and universe appreciation, 151
four phases, 164
and consciousness, 195
and the external world, 200
evolution, 203
disciplining, 203
journey of, 29

phases-journey, 29
to body directives, 207
and sense data, 211-12
and discrimination, 97-98
body complex, 97
Mono-cellular
form, 138
Organism
evolution of, 116
absorbing of materials, 118
Monotony, 229
of existence
and Will of Absolute, 106-108, 142
Movement of energy in limbs, 124
Multiplicity of conscious state, 12
Muscle growing by gymnast
purpose fulfillment, 121
Musical faculties, 27

N

Natural course of our lives, 227
Nature
sound, rhythm, melodies, 152
following, 187
scheme, 181, 190
objects of, 77
New situations
facing, 146
Non-material entity
object as, 69
Notions of outside, 213-15
Nutrition and growth, 134
Nutritional
function of body, 126-27
needs, 149

O

Object, 67
as non-material entity, 69
Organic compound
evolution into, 90, 129, 161
condition maintained, 92
main function, 126
importance in life, 93
Organism
body, 100
life/mind, 100
Organs
freedom/control, 233
of body and location/formation, 128
Our belief, 2223
Our Creator, 221-22
purpose of, 223
Own absence
conceiving by man, 235-36

P

Perception, 11-12
of a being, 227
aspects of, 132
Person, 228
Personality
direction of, 20
Perverted self love, 238
Phase of
discrimination, 165
integration of mind, 165
Phases of mind, 164-65
Philosophical activities, 161-62
Philosophy and religion
domains of, 102

Physical activities
freedom in, 233
Physical entities, 57, 244
Physical objects
world of, 212
Physical sense organs, 27
role of, 28
Physics laws, 216, 218
Plants
evolution, 79
Power of choice, 37
Presentation
Act of, 90
Preservation of life
in an organism, 95
Prime consciousness, 40, 42, 58, 65, 167
energy spent in, 60-61
change in, 45, 58
implicit, 43-46
symbolic representation, 48-49
Will of the Absolute, 65, 72-73
Principle of causality, 49, 72
Principle of Change, 71, 73-74, 84
Principle of consciousness, 99, 179
Principle of differentiation, 82
working, 118-19
Individuation, 90, 92-93, 96
and integration functions, 120
Principle of discrimination, 126
in mine, 96-97
deliberation
and life mind complex functions, 130
Principle of energy, 171
Principle of Freedom of the Will, 142
Principle of individuation, 130-31, 172
Principle of integration, 83, 129
Principle of limitation, 63
Principle of Manifestation of energy, 72-74, 76, 81, 84, 89-90, 99-100,108, 110, 116, 143, 178
laws derived by, 74
Principle of matter, 51
Principle of time, 73
Principle of Will of the Absolute, 74-75, 84, 88-89, 146
Principles and laws
impact on energy, 77
Principles guiding course of life, 92
Purpose-Attempt cycle of body, 120

Q

Qualitative multiplicity, 13
Quality/quantity of food consumed, 135

R

Rational faculties, 235
Real knowledge, 218
acquiring, 238
Relaxing and tension, 26
Religion, 221, 235
Religious
feelings, 221
activities, 161-62, 225
thoughts and actions behind, 231

Reorganising of energy, 126
Resolution, 17
Respect and reverence
  attitudes, 184, 221
  development of, 184
Rhythm, 152

S

Scheme of nature, 191
Schopenhauer, 24, 40
Scientific laws, 77, 216-17
  operating on energy, 73-74, 76
Scientific knowledge, 206
  of a man
    role of, 204
  of the body, 205
Self-existing images, 26
Selfish activities, 189, 238
Self-realization, 230
Selves
  projection of, 18
Sensations
  aspects, 13
Sense data, 27, 208-210
  physical objects represented, 209-211
  compounding, 232
Sense of esteem, 184
  development, 184
Sense of mine and thine, 96
Sense organs, 27
  function of, 137
  growth/development, 150-51
  number in different species, 150
  in amoeba, 151
  forming by consciousness, 196
  role of, 28
Something, 67
Sound in nature, 152
Space, 24, 26, 71, 76, 173-74
  conception of, 169, 174-75
  projecting feelings within, 14
  homogeneity, 151
  in prime consciousness, 43, 69
  and consciousness, 54
  and matter, 54-55 features, 56
Space-time-causality, 198-99
Spatial energy
  and prime consciousness, 69
Spatial order, 56-57
Spirituality, 19
Stages of evolution of body, 138
Status consciousness, 186
Stone as non-living -consciousness aspect, 228
Struggle with
  environment, 179
  own self, 179
Struggles of living beings, 179
Sub-conscious regions of mind, 125
Substitution, 17
Suffering, 239
Suitable environment, 109-110
Superficial
  aspect of mind, 21, 29
  self preference, 16
Symbols/ideas in mind, 199-200

T

Teeth growing in body, 120
  and purpose fulfillment, 120
Tension
  of our mind, 23, 26
  and relaxing, 26, 28
Time, 22, 25, 28, 175
  and homogeneity, 14-15
  concept, 15
  and space, 40-41, 45
Time
  consciousness, 40-42
  thoughts of, 177
Truths
  deductive and inductive, 218

U

Ultimate
  destiny, 236
  goal of life, 189
Unconscious activities, 234
Units of energy, 63-64
Universality in knowledge, 198
Universe, 105
  principle, 141
Unlimited consciousness, 39
Unseen patterns, 229
Urge
  for food, 136
    vs protection of life, 95
  for procreation, 111
Urges and instincts, 136

V

Virtuous activities, 234

W

Wars and conflicts, 183
Water
  evolution, 79
Will
  freedom of, 231
  and deliberation, 61
  exercising, 201
Will of the Absolute, 65, 71-73, 84, 88-89, 104, 171
  or Prime consciousness, 72-73
  manifestation of, 105-106
  and Principles of Change, 110
  principle of, 106
  and monotony of existence, 106-108
Will to determine
  course of action, 52
Work, 59
World of an individual, 212